# Introduction to AutoCAD© 2012

## 2D and 3D Design

1$\frac{1}{16}$"  1$\frac{3}{4}$"  R1$\frac{3}{16}$"

A  T$_A$  1$\frac{13}{16}$"  2"

$\frac{3}{16}$"  7$\frac{1}{2}$"  Ø1$\frac{1}{8}$"

Ø2"  1$\frac{3}{16}$"

3$\frac{13}{16}$"  2"  $\frac{3}{4}$"  $\frac{5}{16}$"  $\frac{3}{4}$"

SECTION A-A

| M. R. AUTOCAD | Scale 1:1 | 15:06:2007 | ANSWER to CHAPTER 8 EXERCISE 1 |

# Alf Yarwood

# Introduction to AutoCAD© 2012

## 2D and 3D Design

**Autodesk®**

Authorized Author

Routledge
Taylor & Francis Group

LONDON AND NEW YORK

First published 2011
by Routledge
2 Park Square, Milton Park, Abingdon, Oxon OX14 4RN

Simultaneously published in the USA and Canada
by Routledge
711 Third Avenue, New York, NY 10017

Routledge is an imprint of the Taylor & Francis Group, an informa business

British Library Cataloguing in Publication Data
A catalogue record for this book is available from the British Library

Library of Congress Number: 2011926755

ISBN: 978-0-08-096947-3

Typeset in Sabon
by MPS Limited, a Macmillan Company, Chennai, India

Printed and bound in Italy

# CONTENTS

# PART B ■ 3D DESIGN

# PART C ■ INTERNET TOOLS AND HELP

# Preface

The purpose of this book is to produce a text suitable for students in Further and/or Higher Education who are required to learn how to use the CAD software package AutoCAD® 2012. Students taking examinations based on computer aided design will find the contents of the book of great assistance. The book is also suitable for those in industry wishing to learn how to construct technical drawings with the aid of AutoCAD 2012. It is also a great asset for those who have used previous releases of AutoCAD or simply wish to update their skills to AutoCAD 2012.

The chapters in Part A – 2D Design, dealing with two-dimensional (2D) drawing, will also be suitable for those wishing to learn how to use AutoCAD LT 2012, the two-dimensional (2D) version of this latest release of AutoCAD.

Many readers using previous releases of AutoCAD will find the book's contents largely suitable for use with those earlier versions; but they will also find that AutoCAD 2012 has many enhancements over previous releases, some of which highlighted in detail in Chapter 21.

The contents of the book are essentially a graded course of work, consisting of chapters which give explanations and examples of methods of construction. Each chapter is followed by exercises, which allows the reader to practice what has been learned and to reinforce understanding of the overall topic of the chapter. The first 11 chapters are concerned with constructing technical drawing in two dimensions (2D). These are followed by chapters detailing the construction of three-dimensional (3D) solid drawings and rendering them. The two final chapters describe the Internet tools of AutoCAD 2012 and the place of AutoCAD in the design process. The book finishes with five appendices – a list of tools with their abbreviations, a list of some of the set variables upon which AutoCAD 2012 is based, a list of the Ribbon panels, a list of the dialogs from which 3D views can be set, and a list of keyboard shortcuts.

AutoCAD 2012 is a very complex computer aided design (CAD) software package. A book of this size cannot possibly cover the complexities of all the methods for constructing 2D and 3D drawings available when working with AutoCAD 2012. However, it is hoped that by the time the reader has worked through the contents of the book, he/she will be sufficiently skilled with methods of producing drawings with the software, be able to go on to more advanced constructions with its use, and have gained an interest in the more advanced possibilities available when using AutoCAD.

*Alf Yarwood*
*Salisbury 2011*

## REGISTERED TRADEMARKS

Autodesk® and AutoCAD® are registered in the US Patent and Trademark Office by Autodesk Inc.

Windows® is a registered trademark of the Microsoft Corporation.

# Autodesk®
## Authorized Author

Alf Yarwood is an Autodesk authorized author and a member of the Autodesk Advanced Developer Network.

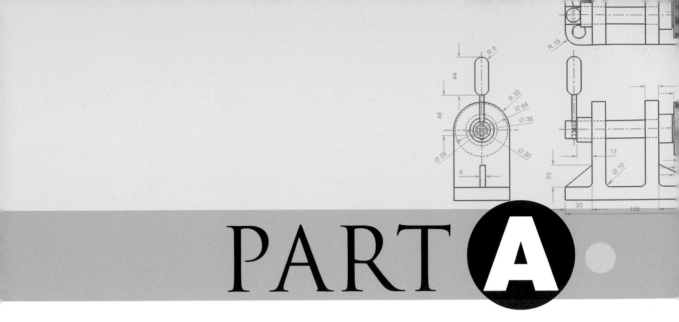

# PART A

## 2D DESIGN

C H A P T E R

# INTRODUCING
# AutoCAD 2012

## AIM OF THIS CHAPTER

The contents of this chapter are designed to introduce features of the AutoCAD 2012 window and methods of operating AutoCAD 2012.

## OPENING AutoCAD 2012

AutoCAD 2012 is designed to work in a Windows operating system. In general, to open AutoCAD 2012 *double-click* on the **AutoCAD 2012** shortcut in the Windows desktop (Fig. 1.1). Depending on how details in **Profiles/Initial Setup...** in the **Options** dialog (Fig. 1.16, page 12) have been set, the **Autodesk Exchange** window (Fig. 1.2) may appear. This dialog allows videos showing methods of working AutoCAD 2012, to be selected from a list of icons.

**Fig. 1.1** The AutoCAD 2012 shortcut in the Windows desktop.

When working in education or in industry, computers may be configured to allow other methods of opening AutoCAD, such as a list appearing on the computer in use when the computer is switched on, from which the operator can select the program he/she wishes to use.

When AutoCAD 2012 is opened a window appears, depending upon whether a **3D Basics**, a **3D Modeling**, a **Classic AutoCAD**, or a **Drafting & Annotation** workspace has been set as **QNEW** (in the **Options** dialog). In this example the **Drafting & Annotation** workspace is shown and includes the **Ribbon** with **Tool** panels (Fig. 1.3). **2D Drafting & Annotation** workspace shows the following details:

> **Ribbon:** which includes tabs, each of which when *clicked* will bring a set of panels containing tool icons. Further tool panels can be seen by *clicking* the appropriate tab. The panels in the ribbon can be changed to any desired panels as required using the **Customize User Interface** dialog if desired (see page 23).

Introduction to AutoCAD 2012. ISBN: 978-0-08-096947-3

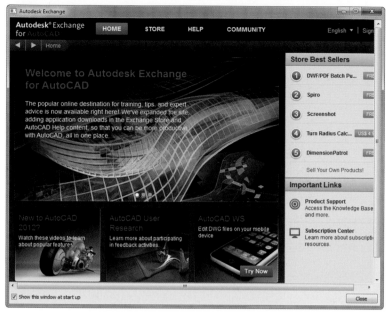

**Fig. 1.2** Page 1 of the **Initial Settings** dialog.

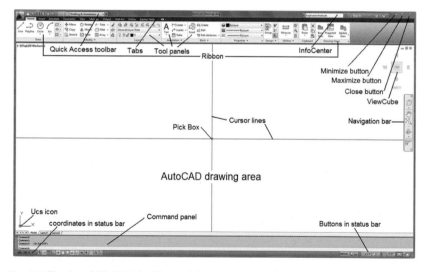

**Fig. 1.3** The AutoCAD 2012 Drafting and Annotation workspace.

**Menu Browser** icon: A *left-click* on the arrow to the right of the **A** symbol at the top left-hand corner of the AutoCAD 2012 window causes the **Menu Browser** menu to appear (Fig. 1.4).

**Workspaces Switching** menu: appears with a *click* on the **Workshop Switching** button in the status bar (Fig. 1.5).

**Command palette:** can be *dragged* from its position at the bottom of the AutoCAD window into the AutoCAD drawing area, when it can be seen to be a palette (Fig 1.6).

**Fig. 1.4** The **Menu Browser.**

**Fig. 1.5** The **Workspace Switching** popup menu.

```
×  Customization file loaded successfully. Customization Group: CONTENTEXPLORER
   Customization file loaded successfully. Customization Group: EXPRESS
   Regenerating model.
   AutoCAD menu utilities loaded.
   Command:
   Command:
```

**Fig. 1.6** The command palette when dragged from its position at the bottom of the AutoCAD window.

**Tool panels:** each shows tools appropriate to the panel. Taking the **Home/Draw** panel as an example, Fig. 1.7 shows that placing the mouse cursor on one of the tool icons in a panel brings a tooltip on screen showing details of how the tool can be used.

**Fig. 1.7** The descriptive tooltip appearing with a click on the **Line** tool icon.

Two types of tooltip will be seen. In the majority of future illustrations of tooltips, the smaller version will be shown. Other tools have popup menus appearing with a *click*. In the example given in Fig. 1.8, a *click* on the **Circle** tool icon will show a

**Fig. 1.8** The tooltip for the **Circle** tool and its popup menu.

tooltip. A *click* on the arrow to the right of the tool icon brings a popup menu showing the construction method options available for the tool.

**Quick Access toolbar:** The toolbar at the top right of the AutoCAD window holds several icons, one of which is the **Open** tool icon. A *click* on the icon opens the **Select File** dialog (Fig. 1.9).

**Navigation bar:** contains several tools which may be of value.

**Fig. 1.9** The **Open** icon in the **Quick Access** toolbar brings the **Select File** dialog to screen.

## THE MOUSE AS A DIGITIZER

Many operators working in AutoCAD will use a two-button mouse as a digitizer. There are other digitizers which may be used – pucks with tablets, a three-button mouse etc. Fig. 1.10 shows a mouse which has two buttons and a wheel.

To operate this mouse, pressing the **Pick button** is a *left-click*. Pressing the **Return** button is a *right-click* which usually, but not always, has the same result as pressing the **Enter** key of the keyboard.

**Fig. 1.10** The two-button mouse.

When the **wheel** is pressed drawings in the AutoCAD screen can be panned by moving the mouse. Moving the wheel forwards enlarges (zooms in) the drawing on screen. Move the wheel backwards and a drawing reduces in size.

The pick box at the intersection of the cursor hairs moves with the cursor hairs in response to movements of the mouse. The AutoCAD window, as shown in Fig. 1.3, shows cursor hairs which stretch across the drawing area in both horizontal and vertical directions. Some operators prefer cursors hairs to be shorter. The length of the cursor hairs can be adjusted in the **Display** sub-menu of the **Options** dialog (page 12).

## PALETTES

A palette has already been shown – the **Command** palette. Two palettes which may be frequently used are the **DesignCenter** palette and the **Properties** palette. These can be called to screen from icons in the **View/Palettes** panel.

> **DesignCenter** palette: Fig. 1.11 shows the **DesignCenter** palette with the **Block** drawings of building symbols from which the block **Third type of chair** has been selected.

**Fig. 1.11** A left-click on the **View/DesignCenter** icon brings the **DesignCenter** palette to screen.

**Properties** palette: Fig. 1.12 shows the **Properties** palette, in which the general features of a selected line are shown. The line can be changed by *entering* new figures in parts of the palette.

**Fig. 1.12** The **Properties** palette.

## TOOL PALETTES

*Click* on **Tool Palettes** in the **View/Palettes** panel and the **Tool Palettes – All Palettes** palette appears (Fig. 1.13).

*Click* in the title bar of the palette and a popup menu appears. *Click* on a name in the menu and the selected palette appears. The palettes can be reduced in size by *dragging* at corners or edges, or hidden by *clicking* on the **Auto-hide** icon, or moved by *dragging* on the **Move** icon. The palette can also be *docked* against either side of the AutoCAD window.

## NOTE →

Throughout this book tools will often be shown as selected from the panels. It will be seen in Chapter 3 that tools can be "called" in a variety of ways, but tools will frequently be shown selected from tool panels although other methods will also be shown on occasion.

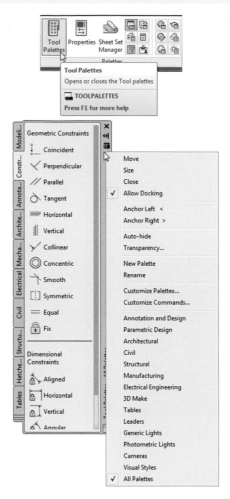

Fig. 1.13 The **Tool Palettes – All Palettes** palette.

## DIALOGS

Dialogs are an important feature of AutoCAD 2012. Settings can be made in many of the dialogs, files can be saved and opened, and changes be made to variables.

Examples of dialogs are shown in Figs 1.15 and 1.16. The first example is taken from the **Select File** dialog (Fig. 1.15), opened with a *click* on **Open...** in the **Quick Access** toolbar (Fig. 1.14). The second example shows part of the **Options** dialog (Fig. 1.16) in which many settings can be made to allow operators the choice of their methods when constructing drawings. The **Options** dialog can be opened with a *click* on **Options...** in the *right-click* menu opened in the command palette.

**Fig. 1.14** Opening the **Select File** dialog from the **Open** icon in the **Quick Access** toolbar.

**Fig. 1.15** The **Select File** dialog.

Note the following parts in the dialog, many of which are common to other AutoCAD dialogs:

**Title bar**: showing the name of the dialog.

**Close dialog button**: common to other dialogs.

**Popup list**: a *left-click* on the arrow to the right of the field brings down a popup list listing selections available in the dialog.

**Buttons**: a *click* on the **Open** button brings the selected drawing on screen. A *click* on the **Cancel** button, closes the dialog.

**Preview** area: available in some dialogs – shows a miniature of the selected drawing or other feature, partly shown in Fig. 1.15.

Note the following in the **Options** dialog (Fig 1.16):

**Tabs**: a *click* on any of the tabs in the dialog brings a sub-dialog on screen.

**Check boxes**: a tick appearing in a check box indicates that the function described against the box is on. No tick means that the

**Fig. 1.16** Part of the **Options** dialog.

function is off. A *click* in a check box toggles between the feature being off or on.

**Radio buttons:** a black dot in a radio button indicates the feature described is on. No dot and the feature is off.

**Slider:** a slider pointer can be *dragged* to change sizes of the feature controlled by the slider.

## BUTTONS AT THE LEFT-HAND END OF THE STATUS BAR

A number of buttons at the left-hand end of the status bar can be used for toggling (turning on/off) various functions when operating within AutoCAD 2012 (Fig. 1.17). A *click* on a button turns that function on, if it is off, a *click* on a button when it is off, turns the function back on. Similar results can be obtained by using the function keys on the computer keyboard (keys **F1** to **F10**).

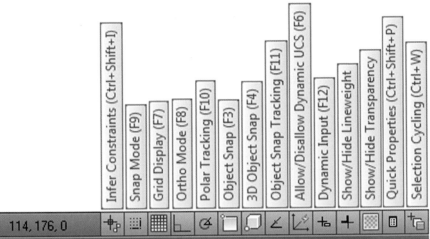

**Fig. 1.17** The buttons at the left-hand end of the status bar.

**Snap Mode:** also toggled using the **F9** key. When snap on, the cursor under mouse control can only be moved in jumps from one snap point to another.

**Grid Display:** also toggled using the **F7** key. When set on, a grid pattern appears in the drawing area.

**Ortho Mode:** also toggled using the **F8** key. When set on, features can only be drawn vertically or horizontally.

**Polar Tracking:** also toggled using the **F10** key. When set on, a small tip appears showing the direction and length of lines etc. in degrees and units.

**Object Snap:** also toggled using the **F3** key. When set on, an **Osnap** icon appears at the cursor pick box.

**Object Snap Tracking:** also toggled by the **F11** key. When set on, lines etc. can be drawn at exact coordinate points and precise angles.

**Allow/Disallow Dynamic UCS:** Dynamic UCS can also be toggled by the **F6** key. Used when constructing 3D solid models.

**Dynamic Input:** Dynamic Input, can also be toggled by **F12.** When set on, the **x,y** coordinates and prompts show when the cursor hairs are moved.

**Show/Hide Lineweight:** when set on, lineweights show on screen. When set off, lineweights only show in plotted/printed drawings.

**Quick Properties:** a *right-click* brings up a popup menu, from which a *click* on **Settings...** causes the **Drafting Settings** dialog to appear.

NOTE →

When constructing drawings in AutoCAD 2012 it is advisable to toggle between **Snap, Ortho, Osnap** and the other functions in order to make constructing easier.

Another set of buttons at the right-hand end of the status bar are shown in Fig. 1.18. The uses of some of these will become apparent when reading future pages of this book. A *click* on the downward facing arrow near the right-hand end of this set of buttons brings up the **Application Status Bar** menu (Fig. 1.19) from which the buttons in the status bar can be set on and/or off.

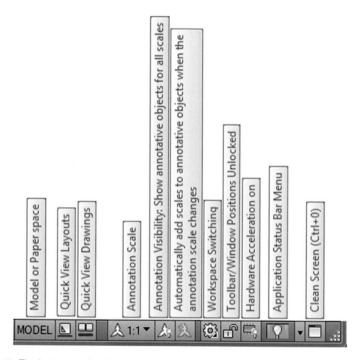

**Fig. 1.18** The buttons at the right-hand end of the status bar.

**Fig. 1.19** The **Application Status Bar** menu.

# THE AutoCAD COORDINATE SYSTEM

In the AutoCAD 2D coordinate system, units are measured horizontally in terms of X and vertically in terms of Y. A 2D point in the AutoCAD drawing area can be determined in terms of X,Y (in this book referred to as $x,y$). $x,y = 0,0$ is the **origin** of the system. The coordinate point $x,y = 100,50$ is 100 units to the right of the origin and 50 units above the origin. The point $x,y = -100,-50$ is 100 units to the left of the origin and 50 points below the origin. Fig. 1.20 shows some 2D coordinate points in the AutoCAD window.

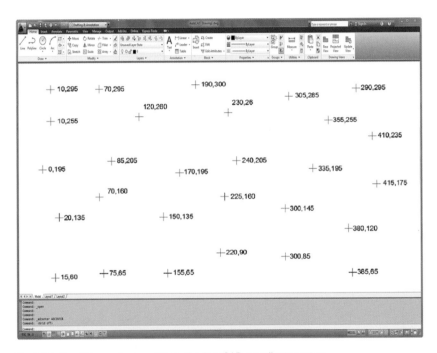

**Fig. 1.20** The 2D coordinate points in the AutoCAD coordinate system.

3D coordinates include a third coordinate (Z), in which positive Z units are towards the operator as if coming out of the monitor screen and negative Z units going away from the operator as if towards the interior of the monitor screen. 3D coordinates are stated in terms of $x,y,z$. $x,y,z = 100,50,50$ is 100 units to the right of the origin, 50 units above the origin, and 50 units towards the operator. A 3D model drawing as if resting on the surface of a monitor is shown in Fig. 1.21.

**Fig. 1.21** A 3D model drawing showing the X, Y, and Z coordinate directions.

## DRAWING TEMPLATES

Drawing templates are files with an extension **.dwt**. Templates are files which have been saved with predetermined settings – such as **Grid** spacing, **Snap** spacing etc... Templates can be opened from the **Select template** dialog (Fig. 1.22) called by *clicking* the **New...** icon in the **Quick Access** toolbar. An example of a template file being opened is shown in Fig. 1.22. In this example the template will be opened in Paper Space and is complete with a title block and borders.

**Fig. 1.22** A template selected from the **Select template** dialog.

When AutoCAD 2012 is used in European countries and opened, the **acadiso.dwt** template is the one most likely to appear on screen. In this part (Part A – 2D Design) of this book drawings will usually be constructed in an adaptation of the **acadiso.dwt** template. To adapt this template:

1. In the command palette *enter* (type) **grid** followed by a *right-click* (or pressing the **Enter** key). Then *enter* **10** in response to the prompt which appears, followed by a *right-click*. (Fig. 1.23).

```
× Command:
  GRID
  Specify grid spacing(X) or [ON/OFF/Snap/Major/aDaptive/Limits/Follow/Aspect]
  <0>: 10

  Command:
```

**Fig. 1.23** Setting **Grids** to **10**.

2. In the command palette *enter* **snap** followed by *right-click*. Then *enter* **5** followed by a *right-click* (Fig. 1.24).

```
× Command:
  Command: SNAP
  Specify snap spacing or [ON/OFF/Aspect/Style/Type] <5>: 5

  Command:
```

**Fig. 1.24** Setting **Snap** to **5**.

3. In the command palette *enter* **limits**, followed by a *right-click*. *Right-click* again. Then *enter* **420,297** and *right-click* (Fig. 1.25).

```
× Command:
  LIMITS
  Reset Model space limits:
  Specify lower left corner or [ON/OFF] <0,0>:
  Specify upper right corner <12,9>: 420,297

  Command:
```

**Fig. 1.25** Setting **Limits** to **420,297**.

4. In the command palette *enter* **zoom** and *right-click*. Then in response to the line of prompts which appears *enter* **a** (for All) and *right-click* (Fig. 1.26).

```
× Command: ZOOM
  Specify corner of window, enter a scale factor (nX or nXP), or
  [All/Center/Dynamic/Extents/Previous/Scale/Window/Object] <real time>: a
  Regenerating model.

  Command:
```

**Fig. 1.26** Zooming to **All**.

5. In the command palette *enter* **units** and *right-click*. The **Drawing Units** dialog appears (Fig. 1.27). In the **Precision** popup list of the **Length** area of the dialog, *click* on 0 and then *click* the **OK** button. Note the change in the coordinate units showing in the status bar.

**Fig. 1.27** Setting **Units** to **0.**

6. *Click* the **Save** icon in the **Quick Access** toolbar (Fig. 1.28). The **Save Drawing As** dialog appears. In the **Files of type** popup list select **AutoCAD Drawing Template (\*.dwt)**. The templates already in AutoCAD are displayed in the dialog. *Click* on **acadiso.dwt,** followed by another *click* on the **Save** button.

**Fig. 1.28** *Click* **Save.**

## NOTES →

1. If in the **Files** area of the **Options** dialog, the **Default template file name for QNEW** is set to **Acadiso.dwt,** when AutoCAD is opened, the template saved as **acadiso.dwt** automatically loads with **Grid** set to **10, Snap** set to **5, Limits** set to **420,297** (size of

→ NOTES (CONTINUED)

an A3 sheet in millimeters) and with the drawing area zoomed to these limits, with **Units** set to **0**.

2. However, if there are multiple users of the computer, it is advisable to save your template to another file name – e.g. **my_template.dwt**.

3. Other features will be added to the template in future chapters.

## METHODS OF SHOWING ENTRIES IN THE COMMAND PALETTE

Throughout this book, when a tool is "called" by a *click* on a tool icon in a panel or, as in this example *entering* **zoom** at the command line, the following will appear in the command palette:

```
Command: enter ZOOM right-click
Specify corner of window, enter a scale factor (nX or
 nXP), or [All/Center/Dynamic/Extents/Previous/Scale/
 Window/Object] <real time>: pick a point on screen
Specify opposite corner: pick another point to form
 a window
Command:
```

## NOTE →

In later examples this may be shortened to:

```
Command: zoom
[prompts]: following by picking points
Command:
```

## NOTES →

1. The above depends on the setting of the variable **SHORCUTMENU**. See page 86

2. In the above *enter* means type the given letter, word or words at the **Command:** prompt.

3. *Right-click* means press the **Return** (right) button of the mouse or press the **Return** key of the keyboard.

## TOOLS AND TOOL ICONS

In AutoCAD 2012, tools are shown as names and icons in panels or in drop-down menus. When the cursor is placed over a tool icon a description shows with the name of the tool as shown and an explanation in diagram form as in the example given in Fig. 1.7.

If a small outward-facing arrow is included at the right-hand side of a tool icon, when the cursor is placed over the icon and the *pick* button of the mouse depressed and held, a flyout appears which includes other features. An example is given in Fig. 1.8 (Page 6).

## ANOTHER AutoCAD WORKSPACE

Other workspaces can be selected as the operator wishes. One in particular which may appeal to some operators is to *click* **AutoCAD Classic** in the **2D Drafting & Annotation** popup menu (Fig. 1.29).

**Fig. 1.29** Selecting **Classic Workspace** from the popup menu.

**Fig. 1.30** The **AutoCAD Classic** workspace.

In the **AutoCAD Classic** workspace (Fig. 1.30), tools icons are held in toolbars, which are *docked* against the sides and top of the workspace. The tool icons in the **Draw** toolbar (*docked* left-hand side) are shown in Fig. 1.31. Note the grid lines, spaced at **10** coordinate units in both **X** and **Y** directions.

## THE RIBBON

In the **2D Drafting & Annotation** workspace, the **Ribbon** contains groups of panels placed at the top of the AutoCAD 2012 window. There are in Fig. 1.3 on page 4, eight panels showing – **Draw**, **Modify**, **Layers**, **Annotation**, **Block**, **Properties**, **Utilities**, and **Clipboard**. Other groups of palettes can be called from the **tabs** at the top of the **Ribbon**.

If a small arrow is showing below the panel name, a *left-click* on the arrow brings down a flyout showing additional tool icons in the panel. As an example Fig. 1.32 shows the flyout from the **Home/Draw** panel.

At the right-hand end of the panel titles (the **tabs**) are two downward pointing arrows. A *left-click* on the right of these two arrows bring down a menu. A *right-click* on the same arrow brings down a different menu (Fig. 1.33). Options from these two menus show that the ribbon can appear in the AutoCAD window in a variety of ways. It is worthwhile experimenting with the settings of the ribbon – each operator will find the best for him or herself. The left-hand arrow also varies the ribbon.

**Fig. 1.31** The tool icons in the Draw Bar.

**Fig. 1.32** The Home/Draw panel and its flyout.

**Fig. 1.33** The menu from the right-hand arrow.

Repeated *left-clicks* on this arrow causes the **Ribbon** panels to:

1. Minimize to tabs;
2. Minimize to panel titles;

3. Minimize to panel button;
4. The full ribbon.

Continuing *clicks* causes the changes to revert to the previous change.

Fig. 1.34 shows the **Minimize** settings. Any one of these settings leaves more space in the AutoCAD drawing window in which to construct drawings. The various settings of the ribbon allow the user discretion as to how to use the ribbon. When minimized to panel titles or to panel buttons, passing the cursor over the titles or buttons causes the panels to reappear and allow selection of tools. Also try **Undock** from the *right-click* menu.

**Fig. 1.34** The **Ribbon** minimize settings.

## THE QUICK VIEW DRAWINGS BUTTON

One of the buttons at the right-hand end of the status bar is the **Quick View Drawings** button. A *click* on this button brings miniatures of recent drawings on screen (Fig. 1.35). This can be of value when wishing to check back features of recent drawings in relation to the current drawing on screen.

**Fig. 1.35** The result of a *click* on the **Quick View Drawings** button.

# CUSTOMIZATION OF THE USER INTERFACE

The AutoCAD 2012 workspace can be arranged in any format the operator wishes by making settings in the **Customize User Interface** dialog (Fig. 1.36) brought to screen from the *right-click* menu, from the button in the **Quick Access** toolbar. The dialog can also be opened using other methods such as *entering* **cui** in the command panel, but using the *right-click* menu is possibly the quickest method. The dialog is only shown here to alert the reader to the fact that he/she can customize the workspace being used to suit their own methods of working. Page space in this book does not allow further explanation.

**Fig. 1.36** The **Customize User Interface** dialog.

## REVISION NOTES ↺

1. A *double-click* on the AutoCAD 2012 shortcut in the Windows desktop opens the AutoCAD window.

2. There are FOUR main workspaces in which drawings can be constructed – the 2D Drafting & Annotation, AutoCAD Classic, 3D Basics, 3D Modeling. Part A – 2D Design of this book deals with 2D drawings will be constructed mainly in the 2D Drafting & Annotation workspace. In Part B – 3D Design, 3D model drawings will be mainly constructed in the 3D Modeling workspace.

3. All constructions in this book involve the use of a mouse as the digitizer. When a mouse is the digitizer:
   - A *left-click* means pressing the left-hand button (the Pick button).
   - A *right-click* means pressing the right-hand button (the Return button).
   - A *double-click* means pressing the left-hand button twice in quick succession.
   - *Dragging* means moving the mouse until the cursor is over an item on screen, holding the left-hand button down and moving the mouse. The item moves in sympathy with the mouse movement.

   To *pick* has a similar meaning to a *left-click*.

4. Palettes are a particular feature of AutoCAD 2012. The Command palette and the DesignCenter palette will be in frequent use.

5. Tools are shown as icons in the tool panels.

6. When a tool is picked, a tooltip describing the tool appears describing the action of the tool. Tools show a small tooltip, followed shortly afterwards by a larger one, but the larger one can be prevented from appearing by selecting an option in the Options dialog.

7. Dialogs allow opening and saving of files and the setting of parameters.

8. A number of *right-click* menus are used in AutoCAD 2012.

9. A number of buttons in the status bar can be used to toggle features such as snap and grid. Functions keys of the keyboard can be also used for toggling some of these functions.

10. The AutoCAD coordinate system determines the position in units of any 2D point in the drawing area (2D Drafting & Annotation) and any point in 3D space (3D Modeling).

11. Drawings are usually constructed in templates with predetermined settings. Some templates include borders and title blocks.

## NOTE →

Throughout this book when tools are selected from panels in the ribbon the tools will be shown in the form for example **Home/ Draw/Line**, the name of the tab in the ribbon title bar, followed by the name of the panel from which the tool is to be selected.

CHAPTER

# INTRODUCING DRAWING

## AIMS OF THIS CHAPTER

The contents of this chapter aim to introduce:

1. The construction of 2D drawing in the **Drafting & Annotation** workspace.
2. The drawing of outlines using the **Line, Circle,** and **Polyline** tools from the **Home/Draw** panel.
3. Drawing to snap points.
4. Drawing to absolute coordinate points.
5. Drawing to relative coordinate points.
6. Drawing using the "tracking" method.
7. The use of the **Erase, Undo,** and **Redo** tools.

## THE DRAFTING & ANNOTATION WORKSPACE

Illustrations throughout this chapter will be shown as working in the **Drafting & Annotation** workspace. In this workspace the **Home/Draw** panel at the left-hand end of the **Ribbon** and **Draw** tools can be selected from the panel as indicated by a *click* on the **Line** tool (Fig. 2.1). In this chapter all examples will show tools as selected from the **Home/Draw** panel. However, methods of construction will be the same if the reader wishes to work by calling tools from the **Draw** drop-down menu. In order to bring drop-down menus on screen, first *click* the small arrow button on the right-hand end of the **Quick Access** toolbar; then *click* **Show Menu Bar** in the menu which appears. Menu titles appear above the **Ribbon**. *Click* **Draw** in this menu bar. From the drop-down menu which appears tools from the **Draw** list in the menu can be selected. Fig. 2.2 shows the **Line** tool being selected.

**Fig. 2.1** The **Line** tool from the **Home/ Draw** panel with its tooltip.

**Fig. 2.2** Selecting the **Line** tool from the **Draw** drop-down menu.

## DRAWING WITH THE LINE TOOL

### FIRST EXAMPLE – LINE TOOL (FIG. 2.4)

1. Open AutoCAD. The drawing area will show the settings of the **acadiso.dwt** template – **Limits** set to **420,297**, **Grid** set to **10**, **Snap** set to **5** and **Units** set to **0**.
2. *Left-click* on the **Line** tool in the **Home/Draw** panel (Fig. 2.1), or *click* **Line** in the **Draw** drop-down menu (Fig. 2.2), or *enter* **line** or **l** at the command line.

## NOTES →

1. The tooltip appears when the tool icon is *clicked* in the **Draw** panel.
2. The prompt **Command: _line Specify first point** appears in the command window at the command line (Fig. 2.3).

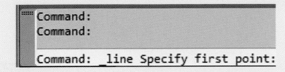

**Fig. 2.3** The prompt which appears at the command line when **Line** is called.

3. Make sure **Snap** is on by either pressing the **F9** key or the **Snap Mode** button in the status bar. **<Snap on>** will show in the command palette.

4. Move the mouse around the drawing area. The cursors pick box will jump from point to point at 5 unit intervals. The position of the pick box will show as coordinate numbers in the status bar (left-hand end).

5. Move the mouse until the coordinate numbers show **60,240,0** and press the *pick* button of the mouse (*left-click*).

6. Move the mouse until the coordinate numbers show **260,240,0** and *left-click*.

7. Move the mouse until the coordinate numbers show **260,110,0** and *left-click*.

8. Move the mouse until the coordinate numbers show **60,110,0** and *left-click*.

9. Move the mouse until the coordinate numbers show **60,240,0** and *left-click*. Then press the **Return** button of the mouse (*right-click*).

    The line rectangle (Fig. 2.4) appears in the drawing area.

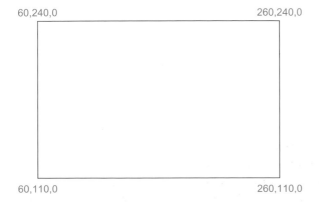

**Fig. 2.4** First example – **Line** tool.

## SECOND EXAMPLE – LINE TOOL (FIG. 2.6)

1. Clear the drawing from the screen with a *click* on the **Close** button of the AutoCAD drawing area. Make sure it is not the AutoCAD 2012 window button.
2. The warning window (Fig. 2.5) appears in the center of the screen. *Click* the **No** button.

**Fig. 2.5** The **AutoCAD** warning window.

3. *Left-click* **New...** button in the **File** drop-down menu and from the **Select template** dialog which appears *double-click* on **acadiso.dwt**.
4. *Left-click* on the **Line** tool icon and *enter* figures as follows at each prompt of the command line sequence:

```
Command: _line Specify first point: enter 80,235
    right-click
Specify next point or [Undo]: enter 275,235
    right-click
Specify next point or [Undo]: enter 295,210
    right-click
Specify next point or [Close/Undo]: enter 295,100
    right-click
Specify next point or [Close/Undo]: enter 230,100
    right-click
Specify next point or [Close/Undo]: enter 230,70
    right-click
Specify next point or [Close/Undo]: enter 120,70
    right-click
Specify next point or [Close/Undo]: enter 120,100
    right-click
Specify next point or [Close/Undo]: enter 55,100
    right-click
Specify next point or [Close/Undo]: enter 55,210
    right-click
Specify next point or [Close/Undo]: enter c (Close)
    right-click
Command:
```

The result is as shown in Fig. 2.6.

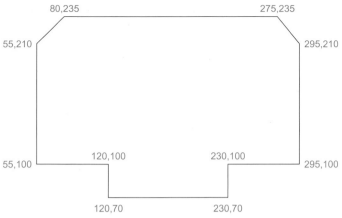

**Fig. 2.6** Second example – **Line** tool.

## THIRD EXAMPLE – LINE TOOL (FIG. 2.7)

1. Close the drawing and open a new **acadiso.dwt** window.
2. *Left-click* on the **Line** tool icon and *enter* figures as follows at each prompt of the command line sequence:

```
Command: _line Specify first point: enter 60,210
   right-click
Specify next point or [Undo]: enter @50,0
   right-click
Specify next point or [Undo]: enter @0,20 right-click
Specify next point or [Undo/Undo]: enter @130,0
   right-click
Specify next point or [Undo/Undo]: enter @0,-20
   right-click
Specify next point or [Undo/Undo]: enter @50,0
   right-click
Specify next point or [Close/Undo]: enter @0,-105
   right-click
Specify next point or [Close/Undo]: enter @-50,0
   right-click
Specify next point or [Close/Undo]: enter @0,-20
   right-click
Specify next point or [Close/Undo]: enter @-130,0
   right-click
Specify next point or [Close/Undo]: enter @0,20
   right-click
Specify next point or [Close/Undo]: enter @-50,0
   right-click
Specify next point or [Close/Undo]: enter c (Close)
   right-click
Command:
```

The result is as shown in Fig. 2.7.

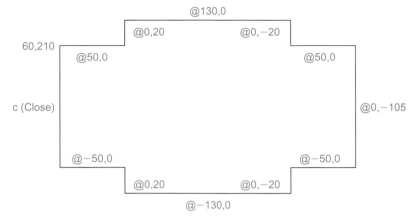

**Fig. 2.7** Third example – **Line** tool.

NOTES →

1. The figures typed at the keyboard determining the corners of the outlines in the above examples are two-dimensional (2D) **x,y** coordinate points. When working in 2D, coordinates are expressed in terms of two numbers separated by a comma.

2. Coordinate points can be shown in positive or negative numbers.

3. The method of constructing an outline as shown in the first two examples above is known as the **absolute coordinate entry** method, where the **x,y** coordinates of each corner of the outlines are *entered* at the command line as required.

4. The method of constructing an outline as in the third example is known as the **relative coordinate entry** method – coordinate points are *entered* relative to the previous entry. In relative coordinate entry, the **@** symbol is *entered* before each set of coordinates with the following rules in mind:

   **+ve** x entry is to the right.

   **−ve** x entry is to the left.

   **+ve** y entry is upwards.

   **−ve** y entry is downwards.

5. The next example (the fourth) shows how lines at angles can be drawn taking advantage of the relative coordinate entry method. Angles in AutoCAD are measured in 360 degrees in a counter-clockwise (anticlockwise) direction (Fig. 2.8). The **<** symbol precedes the angle.

→ NOTES (CONTINUED)

**Fig. 2.8** The counter-clockwise direction of measuring angles in AutoCAD.

## FOURTH EXAMPLE – LINE TOOL (FIG. 2.9)

**1.** Close the drawing and open a new **acadiso.dwt** window.
**2.** *Left-click* on the **Line** tool icon and *enter* figures as follows at each prompt of the command line sequence:

```
Command: _line Specify first point: 70,230
Specify next point: @220,0
Specify next point: @0,-70
Specify next point or [Undo]: @115<225
Specify next point or [Undo]: @-60,0
Specify next point or [Close/Undo]: @115<135
Specify next point or [Close/Undo]: @0,70
Specify next point or [Close/Undo]: c (Close)
Command:
```
The result is as shown in Fig. 2.9.

## FIFTH EXAMPLE – LINE TOOL (FIG. 2.10)

Another method of constructing accurate drawings is by using a method known as **tracking**. When **Line** is in use, as each **Specify next point:** appears at the command line, a *rubber-banded* line

**Fig. 2.9** Fourth example – **Line** tool.

appears from the last point *entered*. *Drag* the rubber-band line in any direction and *enter* a number at the keyboard, followed by a *right-click*. The line is drawn in the *dragged* direction of a length in units equal to the *entered* number.

In this example because all lines are drawn in vertical or horizontal directions, either press the **F8** key or *click* the **ORTHO** button in the status bar which will only allow drawing horizontally or vertically.

1. Close the drawing and open a new **acadiso.dwt** window.
2. *Left-click* on the **Line** tool icon and *enter* figures as follows at each prompt of the command line sequence:

```
Command: _line Specify first point: enter 65,220
    right-click
Specify next point: drag to right enter 240 right-
    click
Specify next point: drag down enter 145 right-click
Specify next point or [Undo]: drag left enter 65
    right-click
Specify next point or [Undo]: drag upwards enter 25
    right-click
Specify next point or [Close/Undo]: drag left enter
    120 right-click
Specify next point or [Close/Undo]: drag upwards
    enter 25 right-click
Specify next point or [Close/Undo]: drag left enter
    55 right-click
Specify next point or [Close/Undo]: c (Close) right-
    click
Command:
```

The result is as shown in Fig. 2.10.

**Fig. 2.10** Fifth example – **Line** tool.

## DRAWING WITH THE CIRCLE TOOL

## FIRST EXAMPLE – CIRCLE TOOL (FIG. 2.13)

1. Close the drawing just completed and open the **acadiso.dwt** template.
2. *Left-click* on the **Circle** tool icon in the **Home/Draw** panel (Fig. 2.11).

**Fig. 2.11** The **Circle** tool from the **Home/Draw** panel.

3. *Enter* a coordinate and a radius against the prompts appearing in the command window as shown in Fig. 2.12, followed by *right-clicks*. The circle (Fig. 2.13) appears on screen.

```
Command:
Command: _circle Specify center point for circle or [3P/2P/Ttr (tan tan
radius)]: 180,150
Specify radius of circle or [Diameter]: 55

Command:
```

**Fig. 2.12** The command line prompts when **Circle** is called.

**Fig. 2.13** First example – **Circle** tool.

# SECOND EXAMPLE – CIRCLE TOOL (FIG. 2.15)

1. Close the drawing and open the **acadiso.dwt** screen.
2. *Left-click* on the **Circle** tool icon and construct two circles as shown in the drawing (Fig. 2.14) in the positions and radii shown in Fig. 2.15.

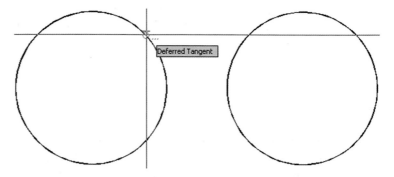

**Fig. 2.14** Second example – **Circle** tool – the two circles of radius 50.

3. *Click* the **Circle** tool again and against the first prompt *enter* **t** (the abbreviation for the prompt **tan tan radius**), followed by a *right-click*.

```
Command _circle Specify center point for circle
    or [3P/2P/Ttr (tan tan radius]: enter t
    right-click
Specify point on object for first tangent of circle:
    pick
Specify point on object for second tangent of
    circle: pick
Specify radius of circle (50): enter 40 right-click
Command:
```

The circle of radius **40** tangential to the two circle already drawn then appears (Fig. 2.15).

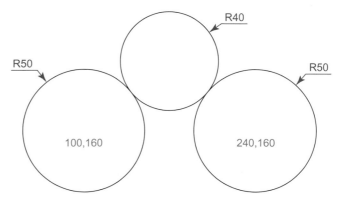

**Fig. 2.15** Second example – **Circle** tool.

## NOTES →

1. When a point on either circle is picked a tip (**Deferred Tangent**) appears. This tip will only appear when the **Object Snap** button is set on with a *click* on its button in the status bar, or the **F3** key of the keyboard is pressed.

2. Circles can be drawn through 3 points or through 2 points *entered* at the command line in response to prompts brought to the command line by using **3P** and **2P** in answer to the circle command line prompts.

## THE ERASE TOOL

If an error has been made when using any of the AutoCAD 2012 tools, the object or objects which have been incorrectly drawn can be deleted using the **Erase** tool. The **Erase** tool icon can be selected from the **Home/Modify** panel (Fig. 2.16) or by *entering* **e** at the command line.

**Fig. 2.16** The **Erase** tool icon from the **Home/Modify** panel.

## FIRST EXAMPLE – ERASE (FIG. 2.18)

**1.** With **Line** construct the outline Fig. 2.17.

**Fig. 2.17** First example – **Erase**. An incorrect outline.

2. Assuming two lines of the outline have been incorrectly drawn, *left-click* the **Erase** tool icon. The command line shows:

```
Command: _erase
Select objects: pick one of the lines 1 found
Select objects: pick the other line 2 total
Select objects: right-click
Command:
```

And the two lines are deleted (right-hand drawing of Fig. 2.18).

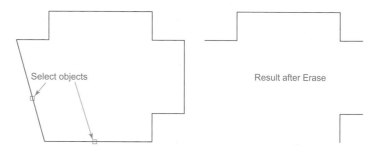

**Fig. 2.18** First example – **Erase.**

## SECOND EXAMPLE – ERASE (FIG. 2.19)

The two lines could also have been deleted by the following method:

1. *Left-click* the **Erase** tool icon. The command line shows:

```
Command: _erase
Select objects: enterc (Crossing)
Specify first corner: pick Specify opposite corner:
  pick 2 found
Select objects: right-click
Command:
```

And the two lines are deleted as in the right-hand drawing (Fig. 2.19).

**Fig. 2.19** Second example – **Erase.**

## UNDO AND REDO TOOLS

Two other tools of value when errors have been made are the **Undo** and **Redo** tools. To undo any last action when constructing a drawing, either *left-click* the **Undo** tool in the **Quick Access** toolbar (Fig. 2.20) or *enter* **u** at the command line. No matter which method is adopted the error is deleted from the drawing.

**Fig. 2.20** The **Undo** tool in the **Quick Access** toolbar.

Everything constructed during a session of drawing can be undone by repeated *clicking* on the **Undo** tool icon or by repeatedly *entering* **u**'s at the command line.

**Fig. 2.21** The **Redo** tool icon in the **Quick Access** toolbar.

To bring back objects that have just been removed by the use of **Undo**'s, *left-click* the **Redo** tool icon in the **Quick Access** toolbar (Fig. 2.21) or *enter* **redo** at the command line.

## DRAWING WITH THE POLYLINE TOOL

When drawing lines with the **Line** tool, each line drawn is an object. A rectangle drawn with the **Line** tool is four objects. A rectangle drawn with the **Polyline** tool is a single object. Lines of different thickness, arcs, arrows and circles can all be drawn using this tool.

Constructions resulting from using the tool are known as **polylines** or **plines**. The tool can be called from the **Home/Draw** panel (Fig. 2.22) or by *entering* **pl** at the command line.

**Fig. 2.22** The **Polyline** tool from **the Home/Draw** panel.

## FIRST EXAMPLE – POLYLINE TOOL (FIG. 2.23)

In this example *enter* and *right-click* have not been included.

*Left-click* the **Polyline** tool. The command line shows:

```
Command: _pline Specify start point: 30,250
Current line width is 0
Specify next point or [Arc/Halfwidth/Length/Undo/
    Width]: 230,250
Specify next point or [Arc/Close/Halfwidth/Length/
    Undo/Width]: 230,120
Specify next point or [Arc/Close/Halfwidth/Length/
    Undo/Width]: 30,120
Specify next point or [Arc/Close/Halfwidth/Length/
    Undo/Width]: c (Close)
Command:
```

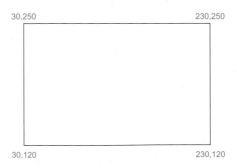

**Fig. 2.23** First example – **Polyline** tool.

NOTE →

1. Note the prompts – **Arc** for constructing pline arcs; **Close** to close an outline; **Halfwidth** to halve the width of a wide pline; **Length** to *enter* the required length of a pline; **Undo** to undo the last pline constructed; and **Width** to change the width of the pline.

2. Only the capital letter(s) of a prompt needs to be *entered* in upper or lower case to make that prompt effective.

3. Other prompts will appear when the **Polyline** tool is in use as will be shown in later examples.

## SECOND EXAMPLE – POLYLINE TOOL (FIG. 2.24)

This will be a long sequence, but it is typical of a reasonably complex drawing using the **Polyline** tool. In the following sequences, when a prompt line is to be repeated, the prompts in square brackets ([]) will be replaced by **[prompts]**.

*Left-click* the **Polyline** tool icon. The command line shows:

```
Command: _pline Specify start point: 40,250
Current line width is 0
Specify next point or [Arc/Halfwidth/Length/Undo/
  Width]: w (Width)
Specify starting width <0>: 5
Specify ending width <5>: right-click
Specify next point or [Arc/Close/Halfwidth/Length/
  Undo/Width]: 160,250
Specify next point or [prompts]: h (Halfwidth)
Specify starting half-width <2.5>: 1
Specify ending half-width <1>: right-click
Specify next point or [prompts]: 260,250
Specify next point or [prompts]: 260,180
Specify next point or [prompts]: w (Width)
Specify starting width <1>: 10
Specify ending width <10>: right-click
Specify next point or [prompts]: 260,120
Specify next point or [prompts]: h (Halfwidth)
Specify starting half-width <5>: 2
Specify ending half-width <2>: right-click
Specify next point or [prompts]: 160,120
Specify next point or [prompts]: w (Width)
Specify starting width <4>: 20
Specify ending width <20>: right-click
Specify next point or [prompts]: 40,120
Specify starting width <20>: 5
Specify ending width <5>: right-click
Specify next point or [prompts]: c (Close)
Command:
```

**Fig. 2.24** Second example – **Polyline** tool.

## THIRD EXAMPLE – POLYLINE TOOL (FIG. 2.25)

*Left-click* the **Polyline** tool icon. The command line shows:
Command: _pline Specify start point: 50,220
Current line width is 0
[prompts]: w (Width)
Specify starting width <0>: 0.5
Specify ending width <0.5>: *right-click*
Specify next point or [prompts]: 120,220
Specify next point or [prompts]: a (Arc)
Specify endpoint of arc or [prompts]: s (second pt)
Specify second point on arc: 150,200
Specify end point of arc: 180,220
Specify end point of arc or [prompts]: l (Line)
Specify next point or [prompts]: 250,220
Specify next point or [prompts]: 260,190
Specify next point or [prompts]: a (Arc)
Specify endpoint of arc or [prompts]: s (second pt)
Specify second point on arc: 240,170
Specify end point of arc: 250,160
Specify end point of arc or [prompts]: l (Line)
Specify next point or [prompts]: 250,150
Specify next point or [prompts]: 250,120
And so on until the outline in Fig. 2.25 is completed.

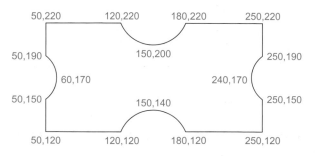

**Fig. 2.25** Third example – **Polyline** tool.

# FOURTH EXAMPLE – POLYLINE TOOL (FIG. 2.26)

*Left-click* the **Polyline** tool icon. The command line shows:

```
Command: _pline Specify start point: 80,170
Current line width is 0
Specify next point or [prompts]: w (Width)
Specify starting width <0>: 1
Specify ending width <1>: right-click
Specify next point or [prompts]: a (Arc)
Specify endpoint of arc or [prompts]: s (second pt)
Specify second point on arc: 160,250
Specify end point of arc: 240,170
Specify end point of arc or [prompts]: cl (CLose)
Command:
```

And the circle in Fig. 2.26 is formed.

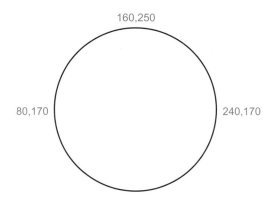

**Fig. 2.26** Fourth example – **Polyline** tool.

# FIFTH EXAMPLE – POLYLINE TOOL (FIG. 2.27)

*Left-click* the **Polyline** tool icon. The command line shows:

```
Command: _pline Specify start point: 60,180
Current line width is 0
Specify next point or [prompts]: w (Width)
Specify starting width <0>: 1
Specify ending width <1>: right-click
Specify next point or [prompts]: 190,180
Specify next point or [prompts]: w (Width)
Specify starting width <1>: 20
Specify ending width <20>: 0
Specify next point or [prompts]: 265,180
Specify next point or [prompts]: right-click
Command:
```

And the arrow in Fig. 2.27 is formed.

60,180     190,180     265,180
Width = 1     Width = 20     Width = 0

**Fig. 2.27** Fifth example – **Polyline** tool.

## REVISION NOTES

**1.** The following terms have been used in this chapter:

*Left-click*: press the left-hand button of the mouse.

*Click*: same meaning as *left-click*.

*Double-click*: press the left-hand button of the mouse twice.

*Right-click*: press the right-hand button of the mouse – has the same result as pressing the Return key of the keyboard.

*Drag*: move the cursor on to a feature and, holding down the left-hand button of the mouse pull the object to a new position. Only applies to features such as dialogs and palettes, not to parts of drawings.

*Enter*: type the letters of numbers which follow at the keyboard.

*Pick*: move the cursor on to an item on screen and press the left-hand button of the mouse.

*Return*: press the *Enter* key of the keyboard. This key may also marked with a left facing arrow. In most cases (but not always) has the same result as a *right-click*.

*Dialog*: a window appearing in the AutoCAD window in which settings may be made.

*Drop-down menu*: a menu appearing when one of the names in the menu bar is *clicked*.

*Tooltip*: the name of a tool appearing when the cursor is placed over a tool icon.

*Prompts*: text appearing in the command window when a tool is selected which advise the operator as to which operation is required.

**2.** Three methods of coordinate entry have been used in this chapter:

*Absolute method*: the coordinates of points on an outline are *entered* at the command line in response to prompts.

*Relative method*: the distances in coordinate units are *entered* preceded by @ from the last point which has been determined on an outline. Angles which are measured in a counter-clockwise direction are preceded by <.

*Tracking*: the rubber band of the line is *dragged* in the direction in which the line is to be drawn and its distance in units is *entered* at the command line followed by a *right-click*.

*Line and Polyline tools*: an outline drawn using the Line tool consists of a number of objects – the number of lines in the outline. An outline drawn using the Polyline is a single object.

## EXERCISES

Methods of constructing answers to the following exercises can be found in the free website:

**http://www.routledge.com/cw/yarwood**

**1.** Using the **Line** tool construct the rectangle in Fig. 2.28.

**2.** Construct the outline in Fig. 2.29 using the **Line** tool. The coordinate points of each corner of the rectangle will need to be calculated from the lengths of the lines between the corners.

**Fig. 2.28** Exercise 1.

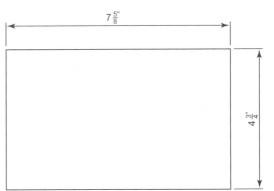

**Fig. 2.29** Exercise 2.

**3.** Using the **Line** tool, construct the outline in Fig. 2.30.

**4.** Using the **Circle** tool, construct the two circles of radius 50 and 30. Then using the **Ttr** prompt add the circle of radius 25 (Fig. 2.31) add the circle of radius 25.

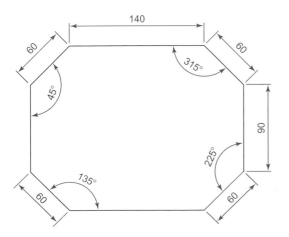

**Fig. 2.30** Exercise 3.

**Fig. 2.31** Exercise 4.

2 **PART A** 2D Design EXERCISES

5. In an **acadiso.dwt** screen and using the **Circle** and **line** tools, construct the line and circle of radius 40 shown in Fig. 2.32. Then, using the **Ttr** prompt, add the circle of radius **25**.

6. Using the **Line** tool construct the two lines at the length and angle as given in Fig. 2.33. Then, with the **Ttr** prompt of the **Circle** tool, add the circle as shown.

**Fig. 2.32** Exercise 5.

**Fig. 2.33** Exercise 6.

7. Using the **Polyline** tool, construct the outline given in Fig. 2.34.

8. Construct the outline in Fig. 2.35 using the **Polyline** tool.

**Fig. 2.34** Exercise 7.

**Fig. 2.35** Exercise 8.

9. With the **Polyline** tool construct the arrows shown in Fig. 2.36.

**Fig. 2.36** Exercise 9.

# DRAW TOOLS, OBJECT SNAP, AND DYNAMIC INPUT

## AIMS OF THIS CHAPTER

The aims of this chapter are:

1. To give examples of the use of the **Arc, Ellipse, Polygon**, and **Rectangle** tools from the **Home/Draw** panel.
2. To give examples of the saving of drawings.
3. To give examples of the uses of the **Polyline Edit** (pedit) tool.
4. To introduce the **Object Snaps** (**Osnaps**) and their uses.
5. To introduce the **Dynamic Input** (**DYN**) system and its uses.

## INTRODUCTION

The majority of tools in AutoCAD 2012 can be called into use by any one of the following six methods:

1. By *clicking* on the tool's icon in the appropriate panel. Fig. 3.1 shows the **Polygon** tool called from the **Home/Draw** panel.
2. By *clicking* on a tool icon in a drop-down menu. Fig. 3.2 shows the tool names and icons displayed in the **Draw** drop-down menu. It is necessary to first bring the menu bar to screen with a *click* on **Show Menu Bar** in the *left-click* menu of the **Quick Access** toolbar (Fig. 3.3) if the menu bar is not already on screen.

Fig. 3.1 The **Polygon** tool and its tooltip selected from the **Home/Draw** panel.

Fig. 3.2 The tool icons in the **Draw** toolbar.

Fig. 3.3 Selecting **Show Menu Bar** from the left-click menu in the **Quick Access** toolbar.

3. By *entering* an abbreviation for the tool name at the command line. For example the abbreviation for the **Line** tool is **l**, for the **Polyline** tool it is **pl**, and for the **Circle** tool it is **c**.

4. By *entering* the full name of the tool at the command line.

5. By making use of the **Dynamic Input** method of construction.
6. If working in the **AutoCAD Classic** workspace by selection of tools from toolbars.

In practice, operators constructing drawings in AutoCAD 2012 may well use a combination of these six methods.

## THE ARC TOOL

In AutoCAD 2012, arcs can be constructed using any three of the following characteristics of an arc: its **Start** point; a point on the arc (**Second** point); its **Center**; its **End**; its **Radius**; the **Length** of the arc; the **Direction** in which the arc is to be constructed; the **Angle** between lines of the arc.

These characteristics are shown in the menu appearing with a *click* on the arrow to the right of the **Arc** tool icon in the **Home/Draw** panel (Fig. 3.4).

**Fig. 3.4** The **Arc** tool flyout in the **Home/Draw** panel.

To call the **Arc** tool *click* on the flyout of its tool icon in the **Home/ Draw** panel, *click* on **Arc** in the **Draw** drop-down menu, or *enter* a or arc at the command line. In the following examples, initials of prompts will be shown instead of selection from the menu as shown in Fig. 3.5.

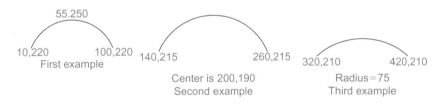

**Fig. 3.5** Examples – **Arc** tool.

## FIRST EXAMPLE – ARC TOOL (FIG. 3.5)

*Left-click* the **Arc** tool icon. The command line shows:

```
Command: _arc Specify start point of arc or
  [Center]: 100,220
Specify second point of arc or [Center/End]: 55,250
Specify end point of arc: 10,220
Command:
```

## SECOND EXAMPLE – ARC TOOL (FIG. 3.5)

```
Command: right-click brings back the Arc sequence
ARC Specify start point of arc or [Center]: c
  (Center)
```

```
Specify center point of arc: 200,190
Specify start point of arc: 260,215
Specify end point of arc or [Angle/chord Length]:
  140,215
Command:
```

## THIRD EXAMPLE – ARC TOOL (FIG. 3.5)

```
Command: right-click brings back the Arc sequence
ARC Specify start point of arc or [Center]:
  420,210
Specify second point of arc or [Center/End]:
  e (End)
Specify end point of arc: 320,210
Specify center point of arc or [Angle/Direction/
  Radius]: r (Radius)
Specify radius of arc: 75
Command:
```

## THE ELLIPSE TOOL

Ellipses can be regarded as what is seen when a circle is viewed from directly in front of the circle and the circle rotated through an angle about its horizontal diameter. Ellipses are measured in terms of two axes – a **major axis** and a **minor axis**, the major axis being the diameter of the circle, the minor axis being the height of the ellipse after the circle has been rotated through an angle (Fig. 3.6).

To call the **Ellipse** tool, *click* on its tool icon in the **Home/Draw** panel (Fig. 3.7), *click* its name in the **Draw** drop-down menu, or *enter* **el** or **ellipse** at the command line.

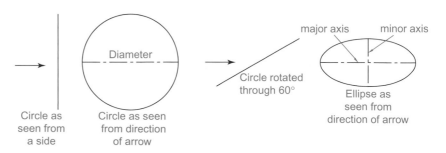

**Fig. 3.6** An ellipse can be regarded as viewing a rotated circle.

**Fig. 3.7** The **Ellipse** tool icon flyout in the **Home/Draw** panel.

## FIRST EXAMPLE – ELLIPSE (FIG. 3.8)

*Left-click* the **Ellipse** tool icon. The command line shows:

```
Command: _ellipse
Specify axis endpoint of elliptical arc or [Center]:
 30,190
Specify other endpoint of axis: 150,190
Specify distance to other axis or [Rotation] 25
Command:
```

## SECOND EXAMPLE – ELLIPSE (FIG. 3.8)

In this second example, the coordinates of the center of the ellipse
(the point where the two axes intersect) are *entered*, followed by
*entering* coordinates for the end of the major axis, followed by
*entering* the units for the end of the minor axis.

```
Command: right-click
ELLIPSE
```

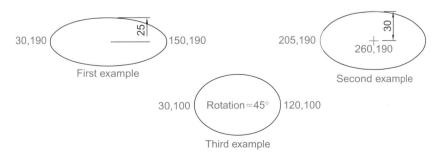

**Fig. 3.8** Examples – **Ellipse.**

```
Specify axis endpoint of elliptical arc or
  [Center]: c
Specify center of ellipse: 260,190
Specify endpoint of axis: 205,190
Specify distance to other axis or
  [Rotation]: 30
Command:
```

## THIRD EXAMPLE – ELLIPSE (FIG. 3.8)

In this third example, after setting the positions of the ends of the major axis, the angle of rotation of the circle from which an ellipse can be obtained is *entered*.

```
Command: right-click
ELLIPSE
Specify axis endpoint of elliptical arc or [Center]:
  30,100
Specify other endpoint of axis: 120,100
Specify distance to other axis or [Rotation]: r
  (Rotation)
Specify rotation around major axis: 45
Command:
```

## SAVING DRAWINGS

Before going further it is as well to know how to save the drawings constructed when answering examples and exercises in this book. Once a drawing has been constructed, *left-click* on **Save As** in the menu appearing with a *left-click* on the AutoCAD icon. This icon appears at the top left-hand corner of the window (Fig. 3.9).

The **Save Drawing As** dialog appears (Fig. 3.10).

Unless you are the only person using the computer on which the drawing has been constructed, it is best to save work to a USB memory stick or other form of temporary saving device. To save a drawing to a USB memory stick:

1. Place a memory stick in a **USB** drive.
2. In the **Save in:** field of the dialog, *click* the arrow to the right of the field and from the popup list select **KINGSTON [F:]** (the name of my **USB** drive and stick).

**Fig. 3.9** Selecting **Save As** from the **Quick Access** menu.

**Fig. 3.10** The **Save Drawing As** dialog.

3. In the **File name:** field of the type a name. The file name extension (**.dwg**) does not need to be typed – it will be added to the file name.

4. *Left-click* the **Save** button of the dialog. The drawing will be saved with the file name extension **.dwg** – the AutoCAD file name extension (Fig. 3.10).

**Fig. 3.11** The Files of type list in the **Save Drawing As** dialog.

NOTES →

In the **Save Drawing As** dialog, *click* in the **Files of type** field. A popup list appears (Fig. 3.11) listing the types of file to which the drawing on screen can be saved:

1. As an AutoCAD *.dwg file in the current version of AutoCAD. Note that the AutoCAD 2010 *.dwg file type has been unchanged for use with AutoCAD 2012.

2. As AutoCAD *.dwg files to be used in earlier versions of AutoCAD. Note that every few years since the first AutoCAD software was released, the format of the AutoCAD *.dwg file has been revised.

→ NOTES (CONTINUED)

3. To any release of AutoCAD LT – the 2D version of AutoCAD.

4. To any release of the AutoCAD or AutoCAD LT *.dxf file type. (See Chapter 10.) *.dxf files can be opened in other types of CAD software.

5. As an AutoCAD template (*.dwt) file.

6. As an AutoCAD Standards (*.dws) file.

   This facility is of value when sharing drawings with others who may be using older versions of AutoCAD.

7. AutoCAD files saved in earlier releases can be opened in AutoCAD 2012 as can AutoCAD LT files.

## SNAP

In previous chapters, several methods of constructing accurate drawings have been described – using **Snap**; absolute coordinate entry; relative coordinate entry, and tracking. Other methods of ensuring accuracy between parts of constructions are by making use of **Object Snaps (Osnaps)**.

**Snap Mode, Grid Display**, and **Object Snaps** can be toggled on/off from the buttons in the status bar or by pressing the keys **F9** (**Snap Mode**), **F7** (**Grid Display**), and **F3** (**Object Snap**).

## OBJECT SNAPS (OSNAPS)

Object Snaps allow objects to be added to a drawing at precise positions in relation to other objects already on screen. With Object Snaps, objects can be added to the end points, mid points, to intersections of objects, to centers and/or quadrants of circles, and so on. Object Snaps also override snap points even when snap is set on.

To set **Object Snaps** – at the command line:

```
Command: enter os
```

And the **Drafting Settings** dialog appears (Fig. 3.12). *Click* the **Object Snap** tab in the upper part of the dialog and *click* the check boxes to the right of the Object Snap names to set them on (or off – check box will then be empty). Fig. 3.12 shows the **Drafting Settings** dialog with some of the **Object Snaps** set on.

**Fig. 3.12** The **Drafting Settings** dialog with some of the **Object Snaps** set on.

When **Object Snaps** are set **on,** as outlines are constructed **Object Snap** icons and their tooltips appear as indicated in Fig. 3.13.

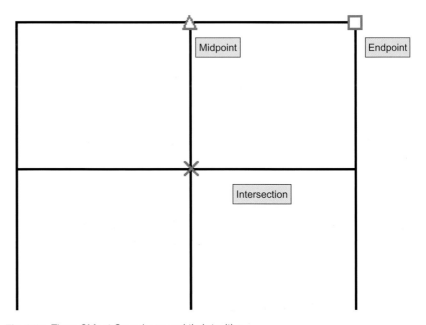

**Fig. 3.13** Three **Object Snap** icons and their tooltips.

It is sometimes advisable not to have **Object Snaps** set on in the **Drafting Settings** dialog, but to set **Object Snap** off and use Object Snap abbreviations at the command line when using tools. The following examples show the use of some of these abbreviations. **Object Snaps** can be toggled on/off by pressing the **F3** key of the keyboard.

## FIRST EXAMPLE – OBJECT SNAP (FIG. 3.14)

Call the **Polyline** tool:

```
Command: _pline
Specify start point: 50,230
[prompts]: w (Width)
Specify starting width: 1
Specify ending width <1>: right-click
Specify next point: 260,230
Specify next point: right-click
Command: right-click
PLINE
Specify start point: pick the right-hand end of the
 pline
Specify next point: 50,120
Specify next point: right-click
Command: right-click
PLINE
Specify start point: pick near the middle of first
 pline
Specify next point: 155,120
Specify next point: right-click
Command: right-click
PLINE
Specify start point: pick the plines at their
 intersection
Specify start point: right-click
Command:
```

The result is shown in Fig. 3.14. In this illustration the **Object Snap** tooltips are shown as they appear when each object is added to the outline.

## SECOND EXAMPLE – OBJECT SNAP ABBREVIATIONS (FIG. 3.15)

Call the **Circle** tool:

```
Command: _circle
Specify center point for circle: 180,170
```

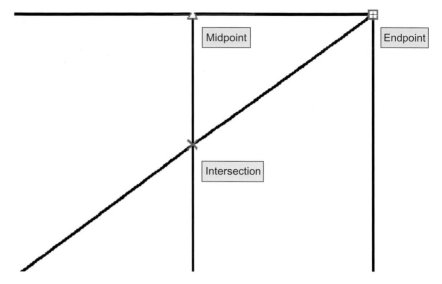

**Fig. 3.14** First example – **Osnaps.**

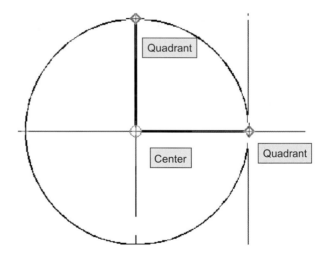

**Fig. 3.15** Second example – **Osnaps.**

```
Specify radius of circle: 60
Command: enter l (Line) right-click
Specify first point: enter qua right-click
of pick near the upper quadrant of the circle
Specify next point: enter cen right-click
of pick near the centre of the circle
Specify next point: enter qua right-click
of pick near right-hand side of circle
Specify next point: right-click
Command:
```

NOTES →

With **Object Snaps** off, the following abbreviations can be used:

**end** – endpoint;
**mid** – midpoint;
**int** – intersection;
**cen** – center;
**qua** – quadrant;
**nea** – nearest;
**ext** – extension.

## DYNAMIC INPUT (DYN)

When **Dynamic Input** is set on either by pressing the **F12** key or with a *click* on the **Dynamic Input** button in the status bar, dimensions, coordinate positions, and commands appear as tips when no tool is in action (Fig. 3.16).

Specify opposite corner: 215 130

**Fig. 3.16** The **DYN** tips appearing when no tool is in action and the cursor is moved.

With a tool in action, as the cursor hairs are moved in response to movement of the mouse, **Dynamic Input** tips showing the coordinate figures for the point of the cursor hairs will show (Fig. 3.17), together with other details. To see the drop-down menu giving the prompts available with **Dynamic Input** press the **down** key of the keyboard and *click* the prompt to be used.

Fig. 3.17 shows the **Arc** prompt as being the next to be used when the **Polyline** tool is in use.

**Fig. 3.17** Coordinate tips when **DYN** is in action.

## NOTES ON THE USE OF DYNAMIC INPUT →

Although **Dynamic Input** can be used in any of the AutoCAD 2012 workspaces, some operators may prefer a larger working area. To achieve this, a *click* on the **Clean Screen** icon in the bottom right-hand corner of the AutoCAD 2012 window produces an uncluttered workspace area. The command palette can be cleared from screen by *entering* **commandlinehide** at the command line. To bring it back press the keys **Ctrl+9**. These two operations produce a screen showing only title and status bars (Fig. 3.18). Some operators may well prefer working in such a larger than normal workspace.

**Dynamic Input** settings are made in the **Dynamic Input** sub-dialog of the **Drafting Settings** dialog (Fig. 3.19), brought to screen by *entering* **os** (or **ds**) at the command line.

When **Dynamic Input** is in action, tools can be called by using any of the methods described on page 57.

1. By *entering* the name of the tool at the command line.
2. By *entering* the abbreviation for a tool name at the command line.
3. By selecting the tool's icon from a panel.
4. By selecting the tool's name from a drop-down menu.

## ➜ NOTES (CONTINUED)

**Fig. 3.18** Example of using **DYN** in a clear screen.

**Fig. 3.19** Settings for **DYN** can be made in the **Drafting Settings** dialog.

When **Dynamic Input** is active and a tool is called, command prompts appear in a tool tip at the cursor position. Fig. 3.20 shows the tool tip appearing at the cursor position when the **Line** tool icon in the **Home/Draw** panel is *clicked*.

To commence drawing a line, either move the cursor under mouse control to the desired coordinate point and *left-click* as in Fig.

→ NOTES (CONTINUED)

**Fig. 3.20** The prompt appearing on screen when the **Line** tool is selected.

3.20, or *enter* the required *x,y* coordinates at the keyboard (Fig. 3.21) and *left-click*. To continue drawing with **Line** *drag* the cursor to a new position and either *left-click* at the position when the coordinates appear as required (Fig. 3.22), or *enter* a required length at the keyboard, which appears in the length box, followed by a *left-click* (Fig. 3.23).

**Fig. 3.21** *Drag* the cursor to the required point and *left-click*.

**Fig. 3.22** *Enter* coordinates for the next point and *left-click*.

**Fig. 3.23** *Enter* length at keyboard and *right-click*.

When using **Dynamic Input** the selection of a prompt can be made by pressing the **down** key of the keyboard (Fig. 3.24), which causes a popup menu to appear. A *click* on the required prompt in such a popup menu will make that prompt active.

# → NOTES (CONTINUED)

The down key of the keyboard

**Fig. 3.24** The **down** key of the keyboard.

## DYNAMIC INPUT – FIRST EXAMPLE – POLYLINE

**1.** Select **Polyline** from the **Home/Draw** panel (Fig. 3.25).

Home

Line | Polyline | Circle | Arc

Draw

Polyline
Creates a 2D polyline

PLINE
Press F1 for more help

**Fig. 3.25** Selecting **Polyline** from the **Home/Draw** panel.

**2.** To start the construction *click* at any point on screen. The prompt for the polyline appears with the coordinates of the selected point showing. *Left-click* to start the drawing (Fig. 3.26).

Specify start point: 90 | 195

**Fig. 3.26 Dynamic Input** – first example – **Polyline** – the first prompt.

**3.** Move the cursor and press the down key of the keyboard. A popup menu appears from which a prompt selection can be made. In the menu *click* **Width** (Fig. 3.27).

**4.** Another prompt field appears. At the keyboard *enter* the required width and *right-click*. Then *left-click* and *enter* **ending width** or *right-click* if the **ending width** is the same as the **starting width** (Fig. 3.28).

**5.** *Drag* the cursor to the right until the dimension shows the required horizontal length and *left-click* (Fig. 3.29).

**Fig. 3.27** **Dynamic Input** – first example – **Polyline** – *click* **Width** in the popup menu.

**Fig. 3.28** **Dynamic Input** – first example – **Polyline** – *entering* widths.

**Fig. 3.29** **Dynamic Input** – first example – **Polyline** – the horizontal length.

**Fig. 3.30** **Dynamic Input** – first example – **Polyline** – the vertical height.

6.  *Drag* the cursor down until the required vertical height is showing and *right-click* (Fig. 3.30).
7.  *Drag* the cursor to the left until the required horizontal distance is showing and *right-click* (Fig. 3.31).
8.  Press the **down** key of the keyboard. *Click* **Close** in the menu (Fig. 3.32). The rectangle completes. Fig. 3.33 shows the completed drawing.

**Fig. 3.31** **Dynamic Input** – first example – **Polyline** – The horizontal distance.

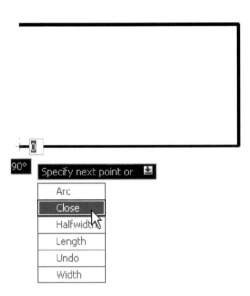

**Fig. 3.32** **Dynamic Input** – first example – **Polyline** – selecting **Close** from the popup menu.

**Fig. 3.33** **Dynamic Input** – first example – **Polyline**.

## DYN – SECOND EXAMPLE – ZOOM

1. *Enter* **Zoom** or **z** at the command line. The first **Zoom** prompt appears (Fig. 3.34).

**Fig. 3.34** **Dynamic Input** – second example – **Zoom** – *enter* **Zoom** at the command line.

2. *Right-click* and press the **down** button of the keyboard. The popup list appears from which a **Zoom** prompt can be selected (Fig. 3.35).

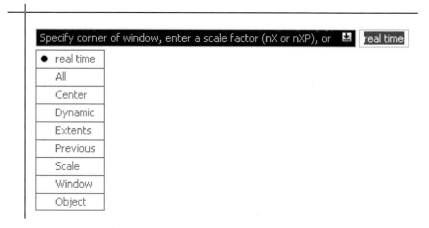

**Fig. 3.35** **Dynamic Input** – second example – **Zoom** – the popup menu appearing with a *right-click* and pressing the **down** keyboard button.

3. Carry on using the **Zoom** tool as described in Chapter 4.

## DYN THIRD EXAMPLE – DIMENSIONING

When using **DYN**, tools can equally as well be selected from a panel. Fig. 3.36 shows the **Linear** tool from the **Home/Annotation** panel selected when dimensioning a drawing.

**Fig. 3.36** Selecting **Linear** from the **Home/Annotation** pane.

A prompt appears asking for the first point. Move the cursor to the second point and another prompt will appear (Fig. 3.37). Press the **down** button of the keyboard and the popup list (Fig. 3.37) appears, from which a selection can be made.

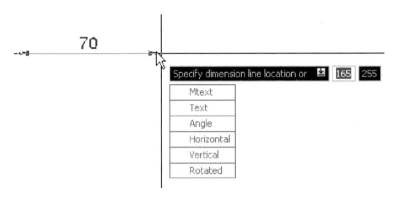

**Fig. 3.37 Dynamic Input –** third example – **dimensioning –** the popup menu associated with **Linear** dimensioning.

The **Dynamic Input** method of constructing 2D drawings can equally as well be used when constructing 3D solid models drawings (See Chapter 12 onwards).

## WHY USE DYNAMIC INPUT?

Some operators may prefer constructing drawings without having to make entries at the command line in response to tool prompts. By using **DYN**, drawings, whether in 2D or in 3D format, can be constructed purely from operating and moving the mouse, *entering* coordinates at the command line and pressing the **down** key of the keyboard when necessary.

## EXAMPLES OF USING OTHER DRAW TOOLS

## POLYGON TOOL (FIG. 3.38)

1. Call the **Polygon** tool – with a *click* on its tool icon in the **Home/Draw** panel (Fig. 3.1, page 46), from the **Draw** drop-down menu, or by *entering* **pol** or **polygon** at the command line. No matter how the tool is called, the command line shows:

```
Command: _polygon Enter number of sides <4>: 6
Specify center of polygon or [Edge]: 60,210
Enter an option [Inscribed in circle/
Circumscribed about circle] <I>: right-click
(accept Inscribed)
Specify radius of circle: 60
Command:
```

2. In the same manner construct a **5**-sided polygon of center **200,210** and of radius **60**.
3. Then, construct an **8**-sided polygon of centre **330,210** and radius **60**.
4. Repeat to construct a **9**-sided polygon circumscribed about a circle of radius **60** and centre **60,80**.
5. Construct yet another polygon with **10** sides of radius **60** and of centre **200,80**.
6. Finally another polygon circumscribing a circle of radius **60**, of centre **330,80** and sides **12**.

The result is shown in Fig. 3.38.

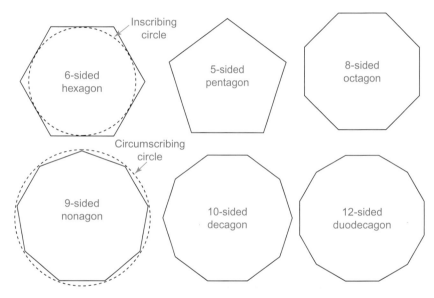

**Fig. 3.38** First example – **Polygon** tool.

## RECTANGLE TOOL – FIRST EXAMPLE (FIG. 3.40)

**Fig. 3.39** The **Rectangle** tool from the **Home/Draw** panel.

Call the **Rectangle** tool – either with a *click* on its tool icon in the **Home/Draw** panel (Fig. 3.39) by *entering* **rec** or **rectangle** at the command line. The tool can be also called from the **Draw** drop-down menu. The command line shows:

```
Command: _rectang
Specify first corner point or [Chamfer/
Elevation/Fillet/Thickness/Width]: 25,240
Specify other corner point or [Area/Dimensions/
 Rotation]: 160,160
Command:
```

## RECTANGLE TOOL – SECOND EXAMPLE (FIG. 3.40)

```
Command: _rectang
[prompts]: c (Chamfer)
Specify first chamfer distance for rectangles <0>: 15
Specify first chamfer distance for rectangles <15>:
 right-click
Specify first corner point: 200,240
Specify other corner point: 300,160
Command:
```

## RECTANGLE TOOL – THIRD EXAMPLE (FIG. 3.40)

```
Command: _rectang
Specify first corner point or [Chamfer/Elevation/
 Fillet/Thickness/Width]: f (Fillet)
Specify fillet radius for rectangles <0>: 15
Specify first corner point or [Chamfer/Elevation/
 Fillet/Thickness/Width]: w (Width)
Specify line width for rectangles <0>: 1
Specify first corner point or [Chamfer/Elevation/
 Fillet/Thickness/Width]: 20,120
Specify other corner point or [Area/Dimensions/
 Rotation]: 160,30
Command:
```

## RECTANGLE – FOURTH EXAMPLE (FIG. 3.40)

```
Command: _rectang
Specify first corner point or [Chamfer/Elevation/
 Fillet/Thickness/Width]: w (Width)
Specify line width for rectangles <0>: 4
```

```
Specify first corner point or [Chamfer/Elevation/
  Fillet/Thickness/Width]: c (Chamfer)
Specify first chamfer distance for rectangles <0>: 15
Specify second chamfer distance for rectangles <15>:
  right-click
Specify first corner point: 200,120
Specify other corner point: 315,25
Command:
```

**Fig. 3.40** Examples – **Rectangle** tool.

## THE POLYLINE EDIT TOOL

The **Polyline Edit** tool is a valuable tool for the editing of polylines.

## FIRST EXAMPLE – POLYLINE EDIT (FIG 3.43)

**1.** With the **Polyline** tool construct the outlines **1** to **6** of Fig. 3.41.

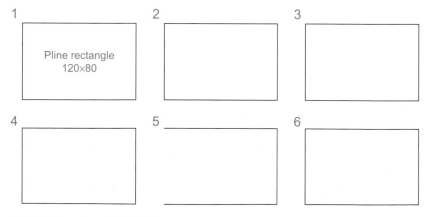

**Fig. 3.41** Examples – **Edit Polyline**.

**2.** Call the **Edit Polyline** tool either from the **Home/Modify** panel (Fig. 3.42) or from the **Modify** drop-down menu, or by *entering* **pe** or **pedit** at the command line, which then shows:

```
Command: enter pe
PEDIT Select polyline or [Multiple]: pick pline 2
 Enter an option [Open/Join/Width/Edit vertex/Fit/
   Spline/Decurve/Ltype gen/Reverse/Undo]: w (Width)
Specify new width for all segments: 2
Enter an option [Open/Join/Width/Edit vertex/Fit/
   Spline/Decurve/Ltype gen/Reverse/Undo]: right-
   click
Command:
```

**Fig. 3.42** Calling **Edit Polyline** from the **Home/Modify** panel.

**3.** Repeat with pline **3** and pedit to Width = **10**.

**4.** Repeat with line **4** and *enter* **s** (Spline) in response to the prompt line:

```
Enter an option [Open/Join/Width/Edit vertex/Fit/
Spline/Decurve/Ltype gen/Reverse/Undo]: enter s
```

**5.** Repeat with pline **5** and *enter* **j** in response to the prompt line:

```
Enter an option [Open/Join/Width/Edit vertex/Fit/
Spline/Decurve/Ltype gen/Undo]: enter j
```

The result is shown in pline **6**.

The resulting examples are shown in Fig. 3.43.

## EXAMPLE – MULTIPLE POLYLINE EDIT (FIG. 3.44)

**1.** With the **Polyline** tool construct the left-hand outlines of Fig. 3.44

**2.** Call the **Edit Polyline** tool. The command line shows:

```
Command: enter pe
PEDIT Select polyline or [Multiple]: m (Multiple)
```

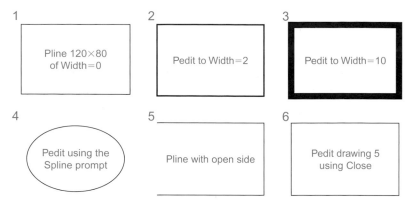

**Fig. 3.43** Examples – **Polyline Edit.**

```
Select objects: pick any one of the lines or arcs
  of the left-hand outlines of Fig. 3.24 found
Select objects: pick another line or arc 1 found
  2 total
Continue selecting lines and arcs as shown by the
  pick boxes of the left-hand drawing of Fig. 3.45
  until the command line shows:
Select objects: pick another line or arc 1 found
  24 total
Select objects: right-click
[prompts]: w (Width)
Specify new width for all segments: 1.5
Convert Arcs, Lines and Splines to polylines
  [Yes/No]? <Y>: right-click
[prompts]: right-click
Command:
```

The result is shown in the right-hand drawing of Fig. 3.44.

**Fig. 3.44** Example – **Multiple Polyline Edit.**

## TRANSPARENT COMMANDS

When any tool is in operation it can be interrupted by prefixing the interrupting command with an apostrophe ('). This is particularly useful when wishing to zoom when constructing a drawing (see page 77). As an example when the **Line** tool is being used:

```
Command: _line
Specify first point: 100,120
Specify next point: 190,120
Specify next point: enter 'z (Zoom)
>> Specify corner of window or [prompts]: pick
>>>>Specify opposite corner: pick
Resuming line command.
Specify next point:
```

And so on. The transparent command method can be used with any tool.

## THE SET VARIABLE PELLIPSE

Many of the operations performed in AutoCAD are carried out under settings of **SET VARIABLES**. Some of the numerous set variables available in AutoCAD 2012 will be described in later pages. The variable **PELLIPSE** controls whether ellipses are drawn as splines or as polylines. It is set as follows:

```
Command: enter pellipse right-click
Enter new value for PELLIPSE <0>: enter 1 right-
  click
Command:
```

And now when ellipses are drawn they are plines. If the variable is set to 0, the ellipses will be splines. The value of changing ellipses to plines is that they can then be edited using the **Polyline Edit** tool.

## REVISION NOTES

1. The following terms have been used in this chapter:
   **Field**: a part of a window or of a dialog in which numbers or letters are *entered* or which can be read.
   **Popup list**: a list brought in screen with a *click* on the arrow often found at the right-hand end of a field.
   **Object**: a part of a drawing which can be treated as a single object.

*Ribbon palettes*: when working in either of the 2D Drafting and Annotation or the 3D Modelling workspace, tool icons are held in panels in the Ribbon.

*Command line*: a line in the command palette which commences with the word Command:

2. *Snap Mode*, *Grid Display*, and *Object Snap* can be toggled with *clicks* on their respective buttons in the status bar. These functions can also be set with function keys: Snap Mode – F9; Grid Display – F7; and Object Snap – F3.

3. Object Snaps ensure accurate positioning of objects in drawings.

4. Object Snap abbreviations can be used at the command line rather than setting in ON in the Drafting Settings dialog.

## NOTES →

There are two types of tooltip. When the cursor under mouse control is placed over a tool icon, the first (a smaller) tooltip is seen. If the cursor is held in position for a short time a second (larger tooltip is seen). Settings for the tooltip may be made in the **Options** dialog.

Polygons constructed with the **Polygon** tool are regular polygons – the edges of the polygons are all the same length and the angles are of the same degrees.

Polygons constructed with the **Polygon** tool are plines, so can be edited by using the **Edit Polyline** tool.

The easiest method of calling the **Edit Polyline** tool is to *enter pe* at the command line.

The **Multiple** prompt of the **pedit** tool saves considerable time when editing a number of objects in a drawing.

Transparent commands can be used to interrupt tools in operation by preceding the interrupting tool name with an apostrophe (').

Ellipses drawn when the variable **PELLIPSE** is set to 0 are splines; when **PELLIPSE** is set to 1, ellipses are polylines. When ellipses are in polyline form they can be modified using the **pedit** tool.

## EXERCISES

Methods of constructing answers to the following exercises can be found in the free website:

**http://www.routledge.com/cw/yarwood**

1. Using the **Line and Arc** tools, construct the outline given in Fig. 3.45.
2. With the **Line** and **Arc** tools, construct the outline in Fig. 3.46.

**Fig. 3.45** Exercise 1.

**Fig. 3.46** Exercise 2.

3. Using the **Ellipse** and **Arc** tools construct the drawing in Fig. 3.47.
4. With the **Line**, **Circle** and **Ellipse** tools construct the drawing in Fig. 3.48.
5. With the **Ellipse** tool, construct the drawing Fig. 3.49.

**Fig. 3.47** Exercise 3.

**Fig. 3.48** Exercise 4.

6. Fig. 3.50 shows a rectangle in the form of a square with hexagons along each edge. Using the **Dimensions** prompt of the **Rectangle** tool construct the square. Then, using the **Edge** prompt of the **Polygon** tool, add the four hexagons. Use the **Object Snap endpoint** to ensure the polygons are in their exact positions.

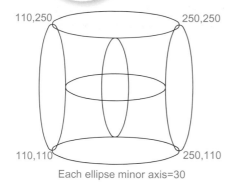

110,250    250,250

110,110    250,110

Each ellipse minor axis=30

**Fig. 3.49** Exercise 5.

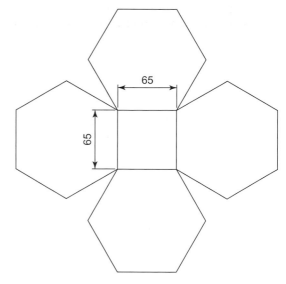

65

65

**Fig. 3.50** Exercise 6.

7. Fig. 3.51 shows seven hexagons with edges touching. Construct the inner hexagon using the **Polygon** tool; then with the aid of the **Edge** prompt of the tool, add the other six hexagons.

8. Fig. 3.52 was constructed using only the **Rectangle** tool. Make an exact copy of the drawing using only the **Rectangle** tool.

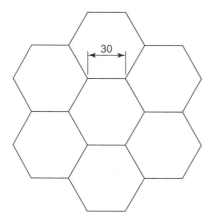

30

**Fig. 3.51** Exercise 7.

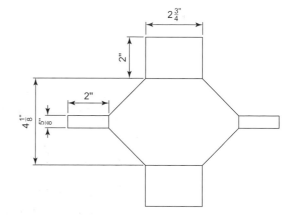

$2\frac{3}{4}$"

2"

2"

$4\frac{1}{8}$"

$\frac{5}{8}$"

**Fig. 3.52** Exercise 8.

9.  Construct the drawing in Fig. 3.53 using the **Line** and **Arc** tools. Then, with the aid of the **Multiple** prompt of the **Edit Polyline** tool change the outlines into plines of **Width = 1**.

**Fig. 3.53** Exercise 9.

10. Construct Fig. 3.54 using the **Line** and **Arc** tools. Then change all widths of lines and arcs to a width of **2** with **Polyline Edit**.

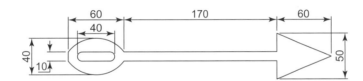

**Fig. 3.54** Exercise 10.

11. Construct Fig. 3.55 using the **Rectangle**, **Line**, and **Edit Polyline** tools.

**Fig. 3.55** Exercise 11.

CHAPTER **4**

# ZOOM, PAN, AND TEMPLATES

## AIMS OF THIS CHAPTER

The aims of this chapter are:

1. To demonstrate the value of the **Zoom** tools.
2. To introduce the **Pan** tool.
3. To describe the value of using the **Aerial View** window in conjunction with the **Zoom** and **Pan** tools.
4. To update the **acadiso.dwt** template
5. To describe the construction and saving of drawing templates.

## INTRODUCTION

The use of the **Zoom** tools allows not only the close inspection of the most minute areas of a drawing in the AutoCAD 2012 drawing area, but allows the accurate construction of very small details in a drawing.

The **Zoom** tools can be called by selection from the **View/Navigate** panel or from the **View** drop-down menu (Fig. 4.1). However, by far the easiest and quickest method of calling the Zoom is to *enter* **z** at the command line as follows:

```
Command: enter z right-click
ZOOM Specify corner of window, enter a scale factor (nX
   or nXP) or [All/Center/Dynamic/Extents/Previous/Scale/
   Window/Object] <real time>:
```

**Introduction to AutoCAD 2012. ISBN: 978-0-08-096947-3**

**Fig. 4.1** Calling **Zooms** – from the **View/Navigate 2D** panel or from the **View** drop-down menu.

This allows the different zooms:

**Realtime:** selects parts of a drawing within a window.

**All:** the screen reverts to the limits of the template.

**Center:** the drawing centers itself around a *picked* point.

**Dynamic:** a broken line surrounds the drawing which can be changed in size and repositioned to part of the drawing.

**Extents:** the drawing fills the AutoCAD drawing area.

**Previous:** the screen reverts to its previous zoom.

**Scale:** entering a number or a decimal fraction scales the drawing.

**Window:** the parts of the drawing within a *picked* window appears on screen. The effect is the same as using **real time**.

**Object:** *pick* any object on screen and the object zooms.

The operator will probably be using **Realtime, Window,** and **Previous** zooms most frequently.

Figure 4.2 shows a drawing that has been constructed.

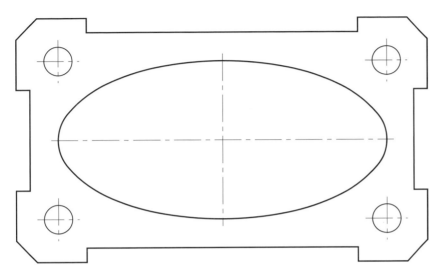

**Fig. 4.2** Drawing to be acted upon by the **Zoom** tool.

Figure 4.3 shows a **Zoom Window** of part of the drawing in Fig. 4.2. This allows it to be checked for accuracy.

**Fig. 4.3** A **Zoom Window** of part of the drawing in Fig. 4.2.

**Fig. 4.4** A **Zoom Extents** of the drawing in Fig. 4.2.

It will be found that the **Zoom** tools are among those most frequently used when working in AutoCAD 2012.

## THE PAN TOOL

The **Pan** tools can be called with a *click* on the **Pan** button in the status bar, from the **Pan** sub-menu of the **View** drop-down menu,

**Fig. 4.5** The drawing used to illustrate Figs 4.6 and 4.7.

or by *entering* **p** at the command line. When the tool is called, the cursor on screen changes to an icon of a hand. *Dragging* the hand across screen under mouse movement allows various parts of a drawing larger than the AutoCAD drawing area to be viewed. As the *dragging* takes place. The **Pan** tool allows any part of the drawing to be viewed and/or modified. When the part of the drawing that is being sought is on screen, a *right-click* calls up the menu from which either the tool can be exited, or other tools can be called. The last part of this process is shown in Fig. 4.6.

**Fig. 4.6** The result of panning Fig. 4.5.

## NOTES →

1. If using a mouse with a wheel both zooms and pans can be performed with the aid of the wheel. See page 7.
2. The **Zoom** tools are important in that they allow even the smallest parts of drawings to be examined and, if necessary, amended or modified.
3. The **Zoom** tools can be called from the sub-menu of the **View** drop-down menu or by *entering* **zoom** or **z** at the command line. The easiest of this choice is to *enter* **z** at the command line followed by a *right-click*.
4. Similarly the easiest method of calling the **Pan** tool is to *enter* **p** at the command line followed by a *right-click*.

## DRAWING TEMPLATES

In Chapters 1 to 3, drawings were constructed in the template **acadiso.dwt** which loaded when AutoCAD 2012 was opened. The default **acadiso** template has been amended to **Limits** set to **420,297** (coordinates within which a drawing can be constructed), **Grid Display** set to **10**, **Snap Mode** set to **5**, and the drawing area **Zoomed** to **All**.

Throughout this book most drawings will be based on an **A3** sheet, which measures 420 units by 297 units (the same as with **Limits**).

### NOTE →

As mentioned before, if others are using the computer on which drawings are being constructed, it is as well to save the template being used to another file name or, if thought necessary, to a memory stick or other temporary type of disk. A file name **My_template.dwt**, as suggested earlier, or a name such as **book_ template** can be given.

## ADDING FEATURES TO THE TEMPLATE

Four other features will now be added to our template:

> **Text style:** set in the **Text Style** dialog.
> **Dimension style:** set in the **Dimension Style Manager** dialog.
> **Shortcutmenu variable:** set to **0**.
> **Layers:** set in the **Layer Properties Manager** dialog.

## SETTING TEXT

**Fig. 4.7** The **Text Style** dialog.

1. At the command line:

```
Command: enter st (Style) right-click
```

2. The **Text Style** dialog appears (Fig. 4.7). In the dialog, *enter* **6** in the **Height** field. Then *left-click* on **Arial** in the **Font Name** popup list. **Arial** font letters appear in the **Preview** area of the dialog.

3. *Left-click* the **New** button and *enter* **Arial** in the **New** text style sub-dialog which appears (Fig. 4.8) and *click* the **OK** button.

**Fig. 4.8** The **New Text Style** sub-dialog.

4. *Left-click* the **Set Current** button of the **Text Style** dialog.
5. *Left-click* the **Close** button of the dialog.

## SETTING DIMENSION STYLE

Settings for dimensions require making *entries* in a number of sub-dialogs in the **Dimension Style Manager.** To set the dimensions style:

1. At the command line:

```
Command: enter d right-click
```

And the **Dimensions Style Manager** dialog appears (Fig. 4.9).

2. In the dialog, *click* the **Modify...** button.
3. The **Modify Dimension Style** dialog appears (Fig. 4.10). This dialog shows a number of tabs at the top of the dialog. *Click* the **Lines** tab and make settings as shown in Fig. 4.10. Then *click* the **OK** button of that dialog.
4. The original **Dimension Style Manager** reappears. *Click* its **Modify** button again.
5. The **Modify Dimension Style** dialog reappears (Fig. 4.11), *click* the **Symbols and Arrows** tab. Set **Arrow** size to 6.

**Fig. 4.9** The **Dimensions Style Manager** dialog.

**Fig. 4.10** The **Modify Dimension Style** dialog.

**Fig. 4.11** Setting **Text** in the **Dimension Style Manager**.

6. Then *click* the **Text** tab. Set **Text style** to **Arial**, set **Color** to **Magenta**, set **Text Height** to **6**, and *click* the **ISO** check box in the bottom right-hand corner of the dialog.

7. Then *click* the **Primary Units** tab and set the units **Precision** to 0 – no units after the decimal point – and **Decimal separator** to **Period**. *Click* the sub-dialogs **OK** button (Fig. 4.12).

**Fig. 4.12** Setting **Primary Units** in the **Dimension Style Manager**.

8. The **Dimension Styles Manager** dialog reappears showing dimensions, as they will appear in a drawing, in the **Preview of the My_Style** box. *Click* the **New...** button. The **Create New Dimension Style** dialog appears (Fig. 4.13).

**Fig. 4.13** The **Create New Dimension Style** dialog.

9. *Enter* a suitable name in the **New style name** field – in this example this is **My_Style**. *Click* the **Continue** button and the **Dimension Style Manager** appears (Fig. 4.14). This dialog now

shows a preview of the **My_Style** dimensions. *Click* the dialog's **Set Current** button, followed by another *click* on the **Close** button. See Fig. 4.14.

**Fig. 4.14** The **Create New Dimension Style** dialog.

## SETTING THE SHORTCUTMENU VARIABLE

**Fig. 4.15** The *right-click* menu.

Call the **Line** tool, draw a few lines, and then *right-click*. The *right-click* menu shown in Fig. 4.15 may well appear. A similar menu will also appear when any tool is called. Some operators prefer using this menu when constructing drawings. To stop this menu appearing:

```
Command: enter shortcutmenu right-click
Enter new value for SHORTCUTMENU <12>: 0
Command:
```

And the menu will no longer appear when a tool is in action.

## SETTING LAYERS

1. At the command line *enter* **layer** or **la** followed by a *right-click*. The **Layer Properties Manager** palette appears (Fig. 4.16).
2. *Click* the **New Layer** icon. **Layer1** appears in the layer list. Overwrite the name **Layer1** *entering* **Center**.
3. Repeat step **2** four times and make four more layers entitled **Construction**, **Dimensions**, **Hidden**, and **Text**.
4. *Click* one of the squares under the **Color** column of the dialog. The **Select Color** dialog appears (Fig. 4.17). *Double-click* on one of the colors in the **Index Color** squares. The selected color appears against the layer name in which the square was selected. Repeat until all 5 new layers have a color.

**Fig. 4.16** The **Layer Properties Manager** palette.

**Fig. 4.17** The **Select Color** dialog.

5. *Click* on the linetype **Continuous** against the layer name **Center**. The **Select Linetype** dialog appears (Fig. 4.18). *Click* its **Load...** button and from the **Load or Reload Linetypes** dialog

**Fig. 4.18** The **Select Linetype** dialog.

*double-click* **CENTER2**. The dialog disappears and the name appears in the **Select Linetype** dialog. *Click* the **OK** button and the linetype **CENTER2** appears against the layer **Center**.

6. Repeat with layer **Hidden**, load the linetype **HIDDEN2** and make the linetype against this layer **HIDDEN2**.

7. *Click* on the any of the lineweights in the **Layer Properties Manager**. This brings up the **Lineweight** dialog (Fig. 4.19). Select the lineweight **0.3**. Repeat the same for all other the layers. Then *click* the **Close** button of the **Layer Properties Manager**.

**Fig. 4.19** The **Lineweight** dialog.

## SAVING THE TEMPLATE FILE

1. *Left-click* on **Save As** in the menu appearing with a *left-click* on the AutoCAD icon at the top left-hand corner of the screen (Fig. 4.20).

2. In the **Save Drawing As** dialog which comes on screen (Fig. 4.21), *click* the arrow to the right of the **Files of type** field and in the popup list associated with the field *click* on **AutoCAD Drawing Template (\*.dwt)**. The list of template files in the **AutoCAD 2012/Template** directory appears in the file list.

3. *Click* on **acadiso** in the file list, followed by a *click* on the **Save** button.

**Fig. 4.20**  Calling **Save As.**

**Fig. 4.21**  Saving the template to the name **acadiso.dwt.**

**4.** The **Template Options** dialog appears. Make *entries* as suggested in Fig. 4.22, making sure that **Metric** is chosen from the popup list.

**Fig. 4.22** The **Template Options** dialog.

The template can now be saved to be opened for the construction of drawings as needed.

Now when AutoCAD 2012 is opened again the template **acadiso. dwt** appears on screen.

NOTE →

Please remember that if others are using the computer it is advisable to save the template to a name of your own choice.

## TEMPLATE FILE TO INCLUDE IMPERIAL DIMENSIONS

If dimensions are to be in **Imperial** measure – in yards, feet, and inches, first set **Limits** to **28,18**. In addition the settings in the **Dimension Style Manager** will need to be different from those shown earlier. Settings for **Imperial measure** in the **Primary Units** sub-dialog need to be set as shown in Fig. 4.23. Settings in the **Text** sub-dialog of the **Text Style** dialog also need to be set as shown in Fig. 4.24.

In addition the settings in the **Primary Units** dialog also need settings to be different to those for Metric dimensions as shown in Fig. 4.25.

**Fig. 4.23** Settings for Imperial dimensions in **Primary Units**.

**Fig. 4.24** Settings for Imperial dimensions set in **Text**.

**Fig. 4.25** Settings for Imperial dimensions in the **Drawing Units** dialog.

## REVISION NOTES

1. The **Zoom** tools are important in that they allow even the smallest parts of drawings to be examined, amended, or modified.

2. The **Zoom** tools can be called from the sub-menu of the View drop-down menu, or by entering z or zoom at the command line. The easiest is to enter z at the command line.

3. There are five methods of calling tools for use:
   - Selecting a tool icon in a panel from a group of panels in the Ribbon.
   - Entering the name of a tool in full at the command line.
   - Entering an abbreviation for a tool.
   - Selecting a tool from a drop-down menu.

   If working in the AutoCAD Classic workspace, tools are called from toolbars.

4. When constructing large drawings, the **Pan tool** and the **Aerial View** window allow work to be carried out in any part of a drawing, while showing the whole drawing in the Aerial View window.

5. An A3 sheet of paper is 420mm by 297mm. If a drawing constructed in the template acadiso.dwt described in this book is printed/plotted full size (scale 1:1), each unit in the drawing will be 1mm in the print/plot.

6. When limits are set it is essential to call Zoom followed by a (All) to ensure that the limits of the drawing area are as set.

7. If the *right-click* menu appears when using tools, the menu can be aborted if required by setting the **SHORTCUTMENU** variable to 0.

## EXERCISES

1. If you have saved drawings constructed either by following the worked examples in this book or by answering exercises in Chapters 2 and 3, open some of them and practice zooms and pans.

2. Go the free website: **http://www.routledge.com/cw/yarwood** and download the drawing given in Chapter 4 Exercise 2 (shown in Fig. 4.26). Open the drawing and practice using the **Zoom** and **Pan** tool to **Zoom** and **Pan** parts of the drawing.

**Fig. 4.26** Exercise 2.

# THE MODIFY TOOLS

## AIM OF THIS CHAPTER

The aim of this chapter is to describe the uses of tools for modifying parts of drawings.

## INTRODUCTION

The **Modify** tools are among those most frequently used. The tools are found in the **Home/Modify** panel. A *click* on the arrow at the bottom of the **Home/Modify** panel brings down a further set of tool icons (Fig. 5.1). They can also be selected from the **Modify** drop-down menu (Fig. 5.2). In the **AutoCAD Classic** workspace, they can be selected from the **Modify** toolbar.

**Fig. 5.1** The **Modify** tool icons in the **Home/Modify** panel.

Using the **Erase** tool from **Home/Modify** was described in Chapter 2. Examples of tools other than the **Explode** follow. See also Chapter 9 for **Explode**.

**Fig. 5.2** The **Modify** drop-down menu.

# THE COPY TOOL

## FIRST EXAMPLE – COPY (FIG. 5.5)

**1.** Construct Fig. 5.3 using **Polyline**. Do not include the dimensions.

**Fig. 5.3** First example – **Copy** – outlines.

**2.** Call the **Copy** tool – either *left-click* on its tool icon in the **Home/Modify** panel (Fig. 5.4), or *enter* **cp** or **copy** at the command line.

**Fig. 5.4** The **Copy** tool from the **Home/Modify** panel.

The command line shows:

```
Command: _copy
Select objects: pick the cross 1 found
Current settings: Copy mode = Multiple
Specify base point or [Displacement/mOde]
   <Displacement>: pick
Specify second point or [Exit/Undo]: pick
Specify second point or [Exit/Undo] <Exit>: right-
   click
Command:
```

The result is given in Fig. 5.5.

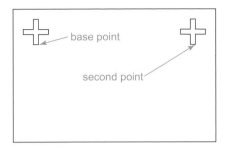

**Fig. 5.5** First example – **Copy.**

# SECOND EXAMPLE – MULTIPLE COPY (FIG. 5.6)

**1.** Erase the copied object.

**2.** Call the **Copy** tool. The command line shows:

```
Command: _copy
Select objects: pick the cross 1 found
Select objects: right-click
Current settings: Copy mode = Multiple
Specify base point or [Displacement/mOde]
   <Displacement>: pick
Specify second point or <use first point as
   displacement>: pick
Specify second point or [Exit/Undo] <Exit>: pick
Specify second point or [Exit/Undo] <Exit>: pick
Command:
```

The result is shown in Fig. 5.6.

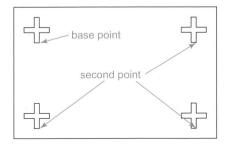

**Fig. 5.6** Second example – **Copy – Multiple copy.**

**Fig. 5.7** First example –
**Mirror** – outline

## THE MIRROR TOOL

### FIRST EXAMPLE – MIRROR (FIG. 5.9)

1.  Construct the outline in Fig. 5.7 using the **Line** and **Arc** tools.
2.  Call the **Mirror** tool – *left-click* on its tool icon in the **Home/Modify** panel (Fig. 5.8) or from the **Modify** drop-down menu, or *enter* **mi** or **mirror** at the command line.

**Fig. 5.8** The **Mirror** tool from the **Home/Modify** panel.

The command line shows:

```
Command: _mirror
Select objects: pick first corner Specify opposite
    corner: pick 7 found
Select objects: right-click
Specify first point of mirror line: pick
Specify second point of mirror line: pick
Erase source objects [Yes/No] <N>: right-click
Command:
```

The result is shown in Fig. 5.9.

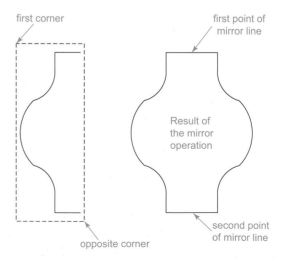

**Fig. 5.9** First example – **Mirror**.

## SECOND EXAMPLE – MIRROR (FIG. 5.10)

1. Construct the outline shown in the dimensioned polyline in the upper drawing of Fig. 5.10.
2. Call **Mirror** and using the tool three times complete the given outline. The two points shown in Fig. 5.10 are to mirror the right-hand side of the outline.

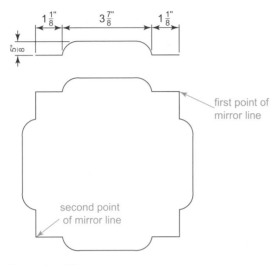

**Fig. 5.10** Second example – **Mirror**.

## THIRD EXAMPLE – MIRROR (FIG. 5.11)

If text is involved when using the **Mirror** tool, the set variable **MIRRTEXT** must be set correctly. To set the variable:

```
Command: mirrtext
Enter new value for MIRRTEXT <1>: 0
Command:
```

If set to **0**, text will mirror without distortion. If set to **1**, text will read backwards as indicated in Fig. 5.11.

**Fig. 5.11** Third example – **Mirror**.

## THE OFFSET TOOL

### EXAMPLES – OFFSET (FIG. 5.14)

1. Construct the four outlines shown in Fig. 5.12.
2. Call the **Offset** tool – *left-click* its tool icon in the **Home/Modify** panel (Fig. 5.12), *pick* the tool name in the **Modify** drop-down menu, or *enter* **o** or **offset** at the command line.

**Fig. 5.12** Examples – **Offset** – outlines.

**Fig. 5.13** The **Offset** tool from the **Home/Modify** panel.

The command line shows:

```
Command: _offset
Current settings: Erase source=No Layer=Source
    OFFSETGAPTYPE=0
Specify offset distance or [Through/Erase/Layer]
    <Through>: 10
Select object to offset or [Exit/Undo] <Exit>: pick
    drawing 1
Specify point on side to offset or [Exit/Multiple/
    Undo] <Exit>: pick inside the rectangle
Select object to offset or [Exit/Undo] <Exit>: e
    (Exit)
Command:
```

3. Repeat for drawings 2, 3, and 4 in Fig. 5.12 as shown in Fig. 5.14.

**Fig. 5.14** Examples – **Offset**.

## THE ARRAY TOOL

Arrays can be either in a **Rectangular** form or in a **Polar** form as shown in the examples below.

## FIRST EXAMPLE – RECTANGULAR ARRAY (FIG. 5.17)

1. Construct the drawing in Fig. 5.15.
2. Call the **Array** tool – either *click* **Array/Rectangular** in the **Modify** drop-down menu (Fig. 5.16), from the **Home/Modify** panel, or *enter* **ar** or **array** at the command line.

**Fig. 5.15** First example – **Array** – drawing to be arrayed.

**Fig. 5.16** Selecting **Rectangular Array** from the **Modify** drop-down menu.

The command line shows:

```
Command: _arrayrect
Select objects: window the drawing 2 found
Select objects: right-click
Type = Rectangular Associative = Yes right-click
Specify base point or enter array type [Rect/PAth/
    POlar/Associative/Key point] <centroid>: pick
    center of drawing
Specify opposite corner for number of items or
    [Angle/Count] <Count>: drag the cursor to the
    right and down until the required array appears
    on screen
Specify opposite corner to space items or [Spacing]
    <Spacing>: right-click
Specify the distance between rows or [Expression]
    <60>: enter -50 right-click
Specify the distance between columns or [Expression]
    <60>: enter 50 right-click
Press Enter to accept or [Base point/Rows/Columns/
    Levels/eXit]<eXit>: right-click
Command:
```

The resulting array is shown in Fig. 5.17.

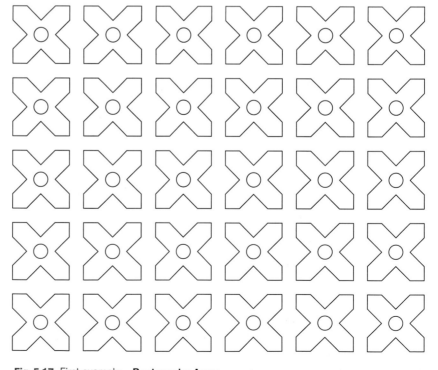

**Fig. 5.17** First example – **Rectangular Array**.

# SECOND EXAMPLE – POLAR ARRAY (FIG. 5.19)

1. Construct the drawing in Fig. 5.18.
2. *Left-click* **Polar Array** in the **Modify** drop-down menu. The command line shows

```
Command: _arraypolar
Select objects: Specify opposite corner: window
    drawing 5 found
Select objects: right-click
Type = Polar Associative = Yes
Specify base point or enter array type [Rect/PAth/
    POlar/Associative/Key point] <centroid>: pick
    center of drawing
Specify center point of array or [Axis of rotation]:
    pick center of array
Enter number of items or [Angle between/Expression]
    <4>: enter 8 right-click
Specify the angle to fill (+=ccw, -=cw) or
    [EXpression] <90> enter 360 right-click
Press Enter to accept or [Base point/Items/Angle
    between/Fill
angle/ROWs/Levels/ROTate items/eXit]<eXit>: right-
    click
Command:
```

The resulting array is shown in Fig 5.19.

**Fig. 5.18** Second example –
the drawing to be arrayed.

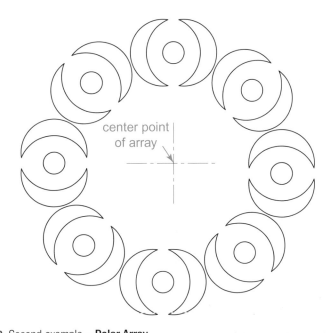

**Fig. 5.19** Second example – **Polar Array**.

## THE MOVE TOOL

### EXAMPLE – MOVE (FIG. 5.22)

**1.** Construct the drawing in Fig. 5.20.

Rectangle 190x50. Chamfers 10x10

All edges are 5

**Fig. 5.20** Example – **Move** – drawing.

**2.** Call **Move** – *click* the **Move** tool icon in the **Home/Modify** panel (Fig. 5.21); *pick* **Move** from the **Modify** drop-down menu.

**Fig. 5.21** The **Move** tool from the **Home/Modify** panel.

Or *enter* **m** or **move** at the command line, which shows:

```
Command: _move
Select objects: pick the middle shape in the drawing
    1 found
Select objects: right-click
Specify base point or [Displacement]
<Displacement>: pick
Specify second point or <use first point as
    displacement>: pick
Command:
```

The result is given in Fig. 5.22.

Fig. 5.22 Example – **Move**.

## THE ROTATE TOOL

When using the **Rotate** tool, remember the default rotation of objects within AutoCAD 2012 is counterclockwise (anticlockwise).

## EXAMPLES – ROTATE (FIG. 5.24)

1. Construct drawing **1** of Fig. 5.24 with **Polyline**. Copy drawing **1** three times (Fig. 5.24).
2. Call **Rotate** – *left-click* its tool icon in the **Home/Modify** panel (Fig. 5.23), *pick* **Rotate** from the **Modify** drop-down menu, or *enter* **ro** or **rotate** at the command line.

Fig. 5.23 The **Rotate** tool from the **Home/Modify** panel.

The command line shows:

```
Command: _rotate
Current positive angle in UCS:
  ANGDIR=counterclockwise ANGBASE=0
Select objects: window the drawing 3 found
Select objects: right-click
Specify base point: pick center of drawing
Specify rotation angle or [Copy/Reference] <0>:
  enter 45 right-click
Command:
```

And the first copy rotates through the specified angle.

3. Repeat for drawings **3** and **4**, rotating as shown in Fig. 5.24.

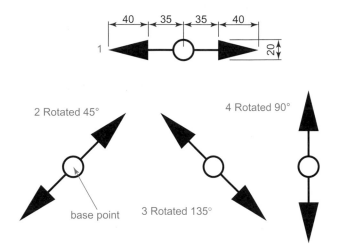

**Fig. 5.24** Examples – **Rotate**.

## THE SCALE TOOL

## EXAMPLES – SCALE (FIG. 5.26)

1. Using the **Rectangle** and **Polyline** tools, construct drawing **1** of Fig. 5.26. The **Rectangle** fillets are R10. The line width of all parts is **1**. Copy the drawing 3 times to give drawings **2**, **3**, and **4**.
2. Call **Scale** – *left-click* its tool icon in the **Home/Modify** panel (Fig. 5.25), *pick* **Scale** from the **Modify** drop-down-menu, or *enter* **sc** or **scale** at the command line which then shows:

**Fig. 5.25** The **Scale** tool from the **Home/Modify** panel.

```
Command: _scale
Select objects: Specify opposite corner: window
   drawing 2 5 found
Select objects: right-click
Specify base point: pick
Specify scale factor or [Copy/Reference]: enter 0.75
   right-click
Command:
```

**3.** Repeat for the other two drawings **3** and **4**, scaling to the scales given with the drawings.

The results are shown in Fig. 5.26.

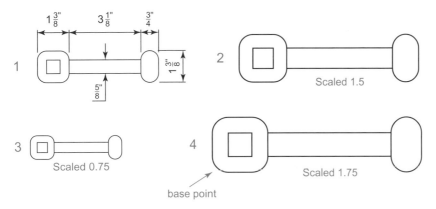

**Fig. 5.26** Examples – **Scale**.

## THE TRIM TOOL

This tool is one which will be in frequent use when constructing drawings.

## FIRST EXAMPLE – TRIM (FIG. 5.28)

**1.** Construct the drawing **Original drawing** in Fig. 5.28 (top).
**2.** Call **Trim** – *left-click* its tool icon in the **Home/Modify** panel (Fig. 5.27);

**Fig. 5.27** The **Trim** tool from the **Home/Modify** panel.

*pick* **Trim** from the **Modify** drop-down menu; or *enter* **tr** or **trim** at the command line, which then shows:

```
Command: _trim
Current settings: Projection UCS. Edge=Extend
Select cutting edges... .
Select objects or <select all>: pick the left-hand
   circle 1 found
Select objects: right-click
Select objects to trim or shift-select to extend or
   [Fence/Project/Crossing/Edge/eRase//Undo]: pick
   one of the objects
Select objects to trim or shift-select to extend or
[Fence/Crossing/Project/Edge/eRase/Undo: pick the
   second of the objects
Select objects to trim or shift-select to extend or
   [Project/Edge/Undo]: right-click
Command:
```

3. This completes the **First stage** as shown in Fig. 5.28. Repeat the **Trim** sequence for the **Second stage**.

**Fig. 5.28** First example – **Trim**.

4. The **Third stage** drawing of Fig. 5.28 shows the result of the trims at the left-hand end of the drawing.
5. Repeat for the right-hand end. The final result is shown in the drawing labelled **Result** in Fig. 5.28.

## SECOND EXAMPLE – TRIM (FIG. 5.29)

1. Construct the left-hand drawing of Fig. 5.29.
2. Call **Trim**. The command line shows:

```
Command: _trim
Current settings: Projection UCS. Edge=Extend
Select cutting edges...
Select objects or <select all>: pick the left-
hand arc 1 found
Select objects: right-click
Select objects to trim or shift-select to extend or
   [Fence/Crossing/Project/Edge/eRase/Undo]: e (Edge)
Enter an implied edge extension mode [Extend/No
   extend] <No extend>: e (Extend)
Select objects to trim: pick
Select objects to trim: pick
Select objects to trim: right-click
Command:
```

3. Repeat for the other required trims. The result is given in Fig. 5.29.

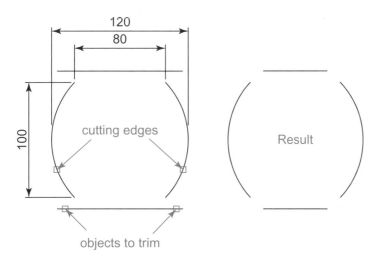

**Fig. 5.29** Second example – **Trim**.

## THE STRETCH TOOL

## EXAMPLES – STRETCH (FIG. 5.31)

As its name implies, the **Stretch** tool is for stretching drawings or
parts of drawings. The action of the tool prevents it from altering

the shape of circles in any way. Only **crossing** or **polygonal** windows can be used to determine the part of a drawing which is to be stretched.

1. Construct the drawing labeled **Original** in Fig. 5.31, but do not include the dimensions. Use the **Circle**, **Arc**, **Trim**, and **Polyline Edit** tools. The resulting outlines are plines of width = 1. With the **Copy** tool make two copies of the drawing.

---

NOTE →

In each of the three examples in Fig. 5.31, the broken lines represent the crossing windows required when **Stretch** is used.

---

2. Call the **Stretch** tool – *click* on its tool icon in the **Home/Modify** panel (Fig. 5.30), *pick* its name in the **Modify** drop-down menu, or *enter* **s** or **stretch** at the command line, which shows.

**Fig. 5.30** The **Stretch** tool from the **Home/Modify** panel.

```
Command: _stretch
Select objects to stretch by crossing-window or
    crossing-polygon...
Select objects: pick a corner of crossing window
    specify opposite corner: 1 found
Select objects: right-click
Specify base point or [Displacement] <Displacement>:
    pick a point in the drawing drag in the direction
    of the stretch
Command:
```

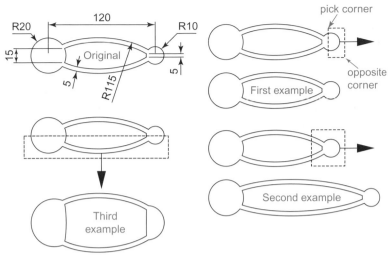

**Fig. 5.31** Examples – **Stretch**.

## NOTES →

1. When circles are windowed with the crossing window no stretching can take place. This is why, in the case of the first example in Fig. 5.33, when the **second point of displacement** was *picked*, there was no result – the outline did not stretch.
2. Care must be taken when using this tool as unwanted stretching can occur.

## THE BREAK TOOL

## EXAMPLES – BREAK (FIG. 5.33)

1. Construct the rectangle, arc, and circle (Fig. 5.33).
2. Call **Break** – *click* its tool icon in the **Home/Modify** panel (Fig. 5.32), *click* **Break** in the **Modify** drop-down menu, or *enter* **br** or **break** at the command line, which shows:

### For Drawings 1 and 2

```
Command: _break Select object: pick at the point
Specify second break point or [First point]: pick
Command:
```

**Fig. 5.32** The **Break** tool icon from the **Home/Modify** panel.

# For Drawing 3

```
Command: _break Select object pick
Specify second break point or [First point]:
enter f right-click
Specify first break point: pick
Specify second break point: pick
Command:
```

The results are shown in Fig. 5.33.

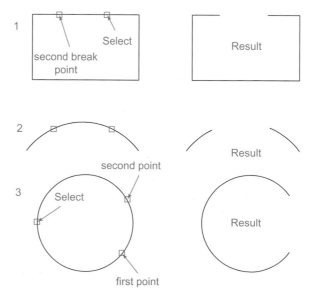

**Fig. 5.33** Examples – **Break**.

## NOTE →

Remember the default rotation of AutoCAD 2012 is counterclockwise. This applies to the use of the **Break** tool.

## THE JOIN TOOL

The **Join** tool can be used to join plines provided their ends are touching, to join lines which are in line with each other, to join arcs, and to convert arcs to circles.

## EXAMPLES – JOIN (FIG. 5.35)

1. Construct a rectangle from four separate plines – drawing **1** of Fig. 5.35; construct two lines – drawing **2** of Fig. 5.35; and construct an arc – drawing **3** of Fig. 5.35.
2. Call the **Join** tool – *click* the **Join** tool icon in the **Home/Modify** panel (Fig. 5.34), select **Join** from the **Modify** drop-down menu, or *enter* **join** or **j**.

**Fig. 5.34** The **Join** tool icon from the **Home/Modify** panel.

```
Command: _join Select source object or multiple
 objects to join at once: pick one side of rectangle
 1 found
Select objects to join: pick second side of
 rectangle 1 found, 2 total
Select objects to join: right-click 2 segments
 joined into 1 polyline
Command: right-click
JOIN Select source object or multiple objects
multiple objects to join at once: pick one arc 1
 found
Select objects to join: pick the second arc 1 found,
 2 total
```

```
Select objects to join: pick an arc 1 found (1
  duplicate), 2 total
Select objects to join: right-click
2 arcs joined into 1 arc
Command: right-click
JOIN Select source object or multiple objects to
  join at once: pick the arc 1 found
Select objects to join:
Select arcs to join to source or [cLose]: enter c
  (cLose)
Arc converted to a circle.
Command:
```

The results are shown in Fig. 5.35.

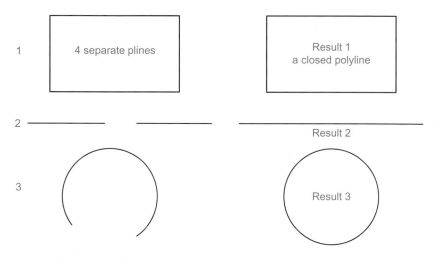

**Fig. 5.35** Examples – **Join**.

## THE EXTEND TOOL

## EXAMPLES – EXTEND (FIG. 5.37)

1. Construct plines and a circle as shown in the left-hand drawings of Fig. 5.37.
2. Call **Extend** – *click* the **Extend** tool icon in the **Home/Modify** panel (Fig. 5.36), *pick* **Extend** from the **Modify** drop-down menu, or *enter* **ex** or **extend** at the command line which then shows:

```
Command: _extend
Current settings: Projection=UCS Edge=Extend
Select boundary edges...
```

**Fig. 5.36** The **Extend** tool icon from the **Home/Modify** panel.

```
Select objects or <select all>: pick the
  vertical line 1 found
Select objects: right-click
Select object to extend or shift-select to trim
  or[Fence/Crossing/Project/Edge/Undo]: pick the
  horizontaL line and the two arcs
Select object to extend or shift-select to trim or
  [Fence/Crossing/Project/Edge/Undo]: right-click
Command:
```

The results are shown in Fig. 5.37.

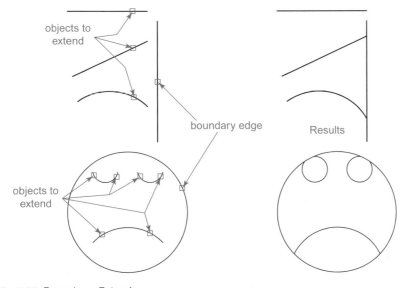

**Fig. 5.37** Examples – **Extend**.

## NOTE →

Observe the similarity of the **Extend** and **No extend** prompts with the **Trim** tool.

## THE FILLET AND CHAMFER TOOLS

These two tools can be called from the **Home/Modify** panel. There are similarities in the prompt sequences for these two tools. The major differences are that only one (**Radius**) setting is required for a fillet, but two settings, **Dist1** and **Dist2**, are required for a chamfer. The basic prompts for each are similar:

### Fillet

```
Command: _fillet
Current settings: Mode=TRIM, Radius=0
Select first object or [Polyline/Radius/Trim/
 mUltiple]: enter r (Radius) right-click
Specify fillet radius <1>: enter 15 right-click
```

### Chamfer

```
Command: _chamfer
(TRIM mode) Current chamfer Dist1=0, Dist2=0
Select first line or [Undo/Polyline/Distance/Angle/
 Trim/mEthod/Multiple]: enter d (Distance) right-
 click
Specify first chamfer distance <0>: enter 10 right-
 click
Specify second chamfer distance <10>: right-click
```

## EXAMPLES – FILLET (FIG. 5.39)

1. Construct three rectangles 100 by 60 using either the **Line** or the **Polyline** tool (Fig.5.39).
2. **Call** Fillet – *click* the arrow to the right of the tool icon in the **Home/Modify** panel and select **Fillet** from the menu which appears (Fig. 5.38), *pick* **Fillet** from the **Modify** drop-down menu,

**Fig. 5.38** Select **Fillet** from the menu in the **Home/Modify** panel.

or *enter* **f** or **fillet** at the command line, which then shows:

```
Command: _fillet
Current settings: Mode=TRIM, Radius=1
Select first object or [Polyline/Radius/Trim/
   Multiple]: enter r (Radius) right-click
Specify fillet radius <0>: enter 15 right-click
Select first object or [Undo/Polyline/Radius/Trim/
   Multiple]: pick
Select second object or shift-select to apply
   corner: pick
Command:
```

Three examples are given in Fig. 5.39.

**Fig. 5.39** Examples – **Fillet.**

# EXAMPLES – CHAMFER (FIG. 5.41)

1. Construct three rectangles 100 by 60 using either the **Line** or the **Polyline** tool.
2. Call **Chamfer** – *click* the arrow to the right of the tool icon in the **Home/Modify** panel and select **Chamfer** from the menu which appears 5.40), *pick* **Chamfer** from the **Modify** drop-down menu,

**Fig. 5.40** Select **Chamfer** from the **Home/Modify** panel.

or *enter* **cha** or **chamfer** at the command line which then shows:

```
Command: _chamfer
(TRIM mode) Current chamfer Dist1=1, Dist2=1
Select first line or [Undo/Polyline/Distance/Angle/
    Trim/
mEthod/Multiple]: d
Specify first chamfer distance <0>: enter 10 right-
    click
Specify second chamfer distance <10>: right-click
Select first line or [Undo/Polyline/Distance/Angle/
    Trim/mEthod/Multiple]: pick the first line for the
    chamfer
Select second line or shift-select to apply corner:
    pick
Command:
```

The result is shown in Fig. 5.41. The other two rectangles are chamfered in a similar manner except that the **No trim** prompt is brought into operation with the bottom left-hand example.

**Fig. 5.41** Examples – **Chamfer.**

## REVISION NOTES ↻

**1.** The Modify tools are among the most frequently used tools in AutoCAD 2012.

**2.** The abbreviations for the Modify tools are:
Copy: cp or co
Mirror: mi
Offset: o
Array: ar
Move: m
Rotate: ro
Scale: sc

## REVISION NOTES (CONTINUED)

Stretch: s
Trim: tr
Extend: ex
Break: br
Join: j
Chamfer: cha
Fillet: f

**3.** There are two other tools in the 2D Draw control panel – Erase – some examples were given in Chapter 2; and Explode – further details of these tools will be given in Chapter 9.

*A note – selection windows and crossing windows*

In the Options dialog settings can be made In the Selection sub-dialog for Visual Effects. A click on the Visual Effects Settings... button brings up another dialog. If the Area Selection Effect settings are set on, a normal window from top left to bottom right will color in a chosen color (default blue). A crossing window – bottom left to top right, will be colored red. Note also that highlighting – select Preview Effect – allows objects to highlight if this feature is on. These settings are shown in Fig. 5.42.

**Fig. 5.42 Visual Effects Settings** sub-dialog of the **Options** dialog.

**4.** When using Mirror, if text is part of the area to be mirrored, the set variable Mirrtext will require setting – to either 1 or 0.

## REVISION NOTES (CONTINUED)

5. With Offset the Through prompt can be answered by clicking two points in the drawing area the distance of the desired offset distance.

6. Polar Arrays can be arrays around any angle set in the Angle of Array field of the Array dialog.

7. When using Scale, it is advisable to practice the Reference prompt.

8. The Trim tool in either its Trim or its No trim modes is among the most useful tools in AutoCAD 2012.

9. When using Stretch circles are unaffected by the stretching.

10. There are some other tools in the Home/Modify panel not described in this book. The reader is invited to experiment with these other tools. They are:

   Bring to Front; Send to Back; Bring above Objects; Send under Objects; Set by Layer; Change Space; Lengthen; Edit Spline; Edit Hatch; Reverse.

EXERCISES

## EXERCISES

Methods of constructing answers to the following exercises can be found in the free website:

**http://www.routledge.com/cw/yarwood**

1.  Construct the diagram in Fig. 5.43. All parts are plines of width = 0.7 with corners filleted R10. The long strips have been constructed using **Circle**, **Polyline**, **Trim**, and **Polyline Edit**. Construct one strip and then copy it using Copy.

2.  Construct the drawing in Fig. 5.44. All parts of the drawing are plines of width = 0.7. The setting in the **Array** dialog is to be **180** in the **Angle of array** field.

**Fig. 5.43** Exercise 1.

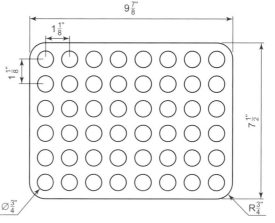

**Fig. 5.44** Exercise 2.

3.  Using the tools **Polyline**, **Circle**, **Trim**, **Polyline Edit**, **Mirror**, and **Fillet** construct the drawing in Fig 5.45.

4.  Construct the circles and lines in Fig. 5.46. Using **Offset** and the **Ttr** prompt of the **Circle** tool followed by **Trim**, construct one of the outlines arrayed within the outer circle. Then, with

**Fig. 5.45** Exercise 3.

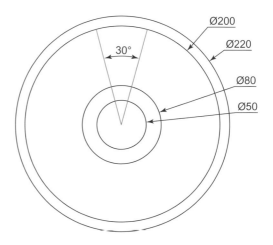

**Fig. 5.46** Exercise 4 – circles and lines on which the exercise is based.

**Polyline Edit** change the lines and arcs into a pline of width = 0.3. Finally, array the outline twelve times around the center of the circles (Fig. 5.47).

5. Construct the arrow in Fig. 5.48. Array the arrow around the center of its circle 8 times to produce the right-hand drawing of Fig. 5.48.

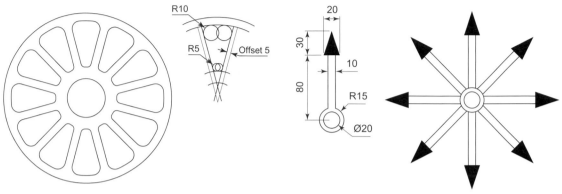

**Fig. 5.47** Exercise 4.

**Fig. 5.48** Exercise 5.

6. Construct the left-hand drawing of Fig. 5.49. Then with **Move**, move the central outline to the top left-hand corner of the outer outline. Then with **Copy** make copies to the other corners.

7. Construct the drawing in Fig. 5.50 and make two copies using **Copy**. With **Rotate**, rotate each of the copies to the angles as shown.

**Fig. 5.49** Exercise 6.

**Fig. 5.50** Exercise 7.

8. Construct the dimensioned drawing in Fig. 5.51. With **Copy**, copy the drawing. Then with **Scale**, scale the drawing to a scale of **0.5**, followed by using **Rotate** to rotate the drawing through an angle as shown. Finally, scale the original drawing to a scale of **2:1**.

**Fig. 5.51** Exercise 8.

9. Construct the left-hand drawing of Fig. 5.52. Include the dimensions in your drawing. Then, using the **Stretch** tool, stretch the drawing, including its dimensions. The dimensions are said to be **associative**.

10. Construct the drawing in Fig. 5.53. All parts of the drawing are plines of width = 0.7. The setting in the **Array** dialog is to be **180** in the **Angle of array** field.

**Fig. 5.52** Exercise 9.

**Fig. 5.53** Exercise 10.

CHAPTER **6**

# DIMENSIONS AND TEXT

## AIMS OF THIS CHAPTER

The aims of this chapter are:
1. To describe a variety of methods of dimensioning drawings.
2. To describe methods of adding text to drawings.

## INTRODUCTION

The dimension style (**My_Style**) has already been set in the **acadiso.dwt** template, which means that dimensions can be added to drawings using this dimension style.

## THE DIMENSION TOOLS

There are several ways in which the dimensions tools can be called.

1. From the **Annotate/Dimensions** panel (Fig. 6.1).
2. *Click* **Dimension** in the menu bar. Dimension tools can be selected from the drop-down menu which appears (Fig. 6.2).
3. By *entering* an abbreviation for a dimension tool at the command line. Some operators may well decide to use a combination of the three methods.
4. In the **Classic AutoCAD** workspace from the **Dimension** toolbar.

**Introduction to AutoCAD 2012. ISBN: 978-0-08-096947-3**

6 **PART A**
2D Design

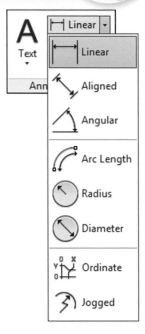

**Fig. 6.1** Dimension tools in the **Annotate/Dimensions** panel.

**Fig. 6.2** Dimensions in the drop-down menu.

NOTE →

In general, in this book dimensions are shown in drawings in the Metric style – mainly in millimeters, but some will be shown in Imperial style – in inches. To see how to set a drawing template for Imperial dimensioning see Chapter 4 (page 77).

## ADDING DIMENSIONS USING THESE TOOLS

### FIRST EXAMPLE – LINEAR DIMENSION (FIG. 6.4)

1. Construct a rectangle 180 × 110 using the **Polyline** tool.
2. Make the **Dimensions** layer current from the **Home/Layers** panel (Fig. 6.3).
3. *Click* the **Linear** tool icon in the **Annotate/Dimensions** panel (Fig. 6.1). The command line shows:

```
Command: _dimlinear
Specify first extension line origin or <select
   object>: pick
```

**Fig. 6.3** The **Home/Layers** panel – making **Dimensions** layer current.

```
Specify second extension line origin: pick
Specify dimension line location or [Mtext/
   Text/Angle/Horizontal/Vertical/Rotated]: pick
Dimension text = 180
Command:
```

Fig. 6.4 shows the 180 dimension. Follow exactly the same procedure for the 110 dimension.

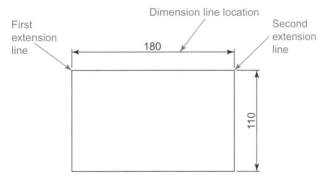

**Fig. 6.4** First example – **Linear** dimension.

## NOTES →

1. If necessary use **Osnaps** to locate the extension line locations.
2. At the prompt

```
Specify first extension line origin or [select
   object]:
```
also allows the line being dimensioned to be *picked*.

→ NOTES (CONTINUED)

3. The drop-down menu from the **Line** tool icon contains the following tool icons – **Angular, Linear, Aligned, Arc Length, Radius, Diameter, Jog Line,** and **Ordinate**. Refer to Fig. 6.1 when working through the examples below. **Note** – when a tool is chosen from this menu, the icon in the panel changes to the selected tool icon.

## SECOND EXAMPLE – ALIGNED DIMENSION (FIG. 6.5)

1. Construct the outline in Fig. 6.5 using the **Line** tool.

**Fig. 6.5** Second example – **Aligned** dimension.

2. Make the **Dimensions** layer current (**Home/Layers** panel).
3. *Left-click* the **Aligned** tool icon (see Fig. 6.1) and dimension the outline. The prompts and replies are similar to the first example.

## THIRD EXAMPLE – RADIUS DIMENSION (FIG. 6.6)

1. Construct the outline in Fig. 6.5 using the **Line** and **Fillet** tools.
2. Make the **Dimensions** layer current (**Home/Layers** panel).
3. *Left-click* the **Radius** tool icon (see Fig. 6.1). The command line shows:

```
Command: _dimradius
Select arc or circle: pick one of the arcs
Dimension text = 30
Specify dimension line location or [Mtext/Text/
  Angle]: pick
Command:
```

4. Continue dimensioning the outline as shown in Fig. 6.6.

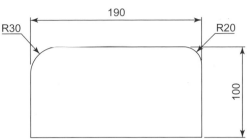

**Fig. 6.6** Third example – **Radius** dimension.

## NOTES →

1. At the prompt
   `[Mtext/Text/Angle]:`
   if a **t** (Text) is *entered*, another number can be *entered*, but remember if the dimension is a radius the letter **R** must be *entered* as a prefix to the new number.
2. If the response is **a** (Angle), and an angle number is *entered* the text for the dimension will appear at an angle.
3. If the response is **m** (Mtext) the **Text Formatting** dialog appears together with a box in which new text can be *entered*. See page 139.
4. Dimensions added to a drawing using other tools from the **Annotate/Dimensions** panel should be practiced.

## ADDING DIMENSIONS FROM THE COMMAND LINE

From Figs 6.1 and 6.2 it will be seen that there are some dimension tools which have not been described in examples. Some operators may prefer *entering* dimensions from the command line. This involves abbreviations for the required dimensions.

For **Linear Dimension: hor** (horizontal) or **ve** (vertical).
For **Aligned Dimension: al**.
For **Radius Dimension: ra**.
For **Diameter Dimension: d**.
For **Angular Dimension: an**.
For **Dimension Text Edit: te**.
For **Quick Leader: l**.

And to exit from the dimension commands: **e** (Exit).

# FIRST EXAMPLE – HOR AND VE (HORIZONTAL AND VERTICAL) – FIG. 6.8

1. Construct the outline in Fig. 6.7 using the **Line** tool. Its dimensions are shown in Fig. 6.8.

**Fig. 6.7** First example – outline to dimension.

2. Make the **Dimensions** layer current (**Home/Layers** panel).
3. At the command line *enter* **dim.** The command line will show:

```
Command: enter dim right-click
Dim: enter hor (horizontal) right-click
Specify first extension line origin or <select
   object>: pick
Specify second extension line origin: pick
Non-associative dimension created.
Specify dimension line location or [Mtext/Text/
   Angle]: pick
Enter dimension text <50>: right-click
Dim: right-click
HORIZONTAL
Specify first extension line origin or <select
   object>: pick
Specify second extension line origin: pick
Non-associative dimension created.
Specify dimension line location or [Mtext/Text/
   Angle/Horizontal/Vertical/Rotated]: pick
Enter dimension text <140>: right-click
Dim: right-click
```

And the 50 and 140 horizontal dimensions are added to the outline.

4. Continue to add the right-hand 50 dimension. Then once the command line shows:

```
Dim: enter ve (vertical) right-click
Specify first extension line origin or <select
   object>: pick
Specify second extension line origin: pick
Specify dimension line location or [Mtext/Text/
   Angle/Horizontal/Vertical/Rotated]: pick
```

```
Dimension text <20>: right-click
Dim: right-click
VERTICAL
Specify first extension line origin or <select
   object>: pick
Specify second extension line origin: pick
Specify dimension line location or [Mtext/Text/
   Angle/Horizontal/Vertical/Rotated]: pick
Dimension text <100>: right-click
Dim: enter e (Exit) right-click
Command:
```

The result is shown in Fig. 6.8.

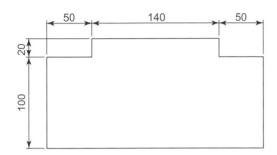

**Fig. 6.8** First example – horizontal and vertical dimensions.

## SECOND EXAMPLE – AN (ANGULAR) – FIG. 6.10

**1.** Construct the outline in Fig. 6.9 – a pline of width = 1.

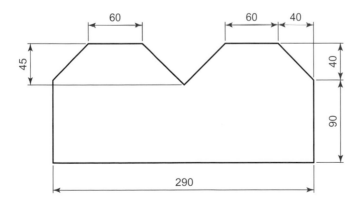

**Fig. 6.9** Second example – outline for dimensions.

**2.** Make the **Dimensions** layer current (**Home/Layers** panel).

**3.** At the command line:

```
Command: enter dim right-click
Dim: enter an right-click
Select arc, circle, line or <specify vertex>: pick
```

```
Select second line: pick
Specify dimension arc line location or [Mtext/Text/
   Angle/Quadrant]: pick
Enter dimension <90>: right-click
Enter text location (or press ENTER): pick
Dim:
```

And so on to add the other angular dimensions.

The result is given in Fig. 6.10.

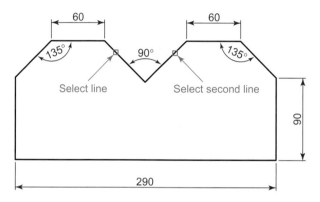

**Fig. 6.10** Second example – **an** (Angle) dimension.

## THIRD EXAMPLE – L (LEADER) – (FIG. 6.12)

**1.** Construct the diagram in Fig. 6.11.

**Fig. 6.11** Third example – outline for dimensioning.

**2.** Make the **Dimensions** layer current (**Home/Layers** panel).

**3.** At the command line:

```
Command: enter dim right-click
Dim: enter l (Leader) right-click
Leader start: enter nea (osnap nearest) right-click
   to pick one of the chamfer lines
To point: or [Undo]: pick
To point: or [Undo]: or [Undo]: pick
```

```
To point: or [Undo]: right-click
Dimension text <0>: enter CHA 10x10 right-click
Dim: right-click
```

Continue to add the other leader dimensions – Fig. 6.12.

**Fig. 6.12** Third example – l (Leader) dimensions.

## FOURTH EXAMPLE – TE (DIMENSION TEXT EDIT) – (FIG. 6.14)

**1.** Construct the diagram in Fig. 6.13.

**Fig. 6.13** Fourth example – dimensioned drawing.

**2.** Make the **Dimensions** layer current (**Home/Layers** panel).

**3.** At the command line:

```
Command: enter dim right-click
Dim: enter te (tedit) right-click
Select dimension: pick the dimension to be changed
Specify new location for text or [Left/Right/Center/
   Home/Angle]: either drag the dimension to one end
   of the dimension line
Dim:
```

The results as given in Fig. 6.14 show dimensions which have been moved, the **210** dimension changed to the left-hand end of the dimension line, the **130** dimension changed to the left-hand end of the dimension line, and the **30** dimension position changed.

**Fig. 6.14** Fourth example – dimensions amended with **tedit**.

## THE ARC LENGTH TOOL (FIG. 6.15)

**1.** Construct two arcs of different sizes as in Fig. 6.15.

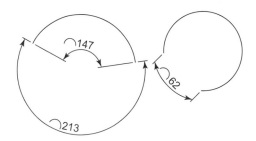

**Fig. 6.15** Examples – **Arc Length** tool.

**2.** Make the **Dimensions** layer current (**Home/Layers** panel).
**3.** Call the **Arc Length** tool from the **Annotate/Dimensions** panel (see Fig. 6.1), or *enter* **dimarc** at the command line. The command line shows:

```
Command: _dimarc
Select arc or polyline arc segment: pick an arc
Specify arc length dimension location, or [Mtext/
    Text/Angle/Partial/Leader]: pick a suitable
    position
Dimension text = 147
Command:
```

## THE JOGGED TOOL (FIG. 6.16)

**1.** Draw a circle and an arc as indicated in Fig. 6.16.
**2.** Make the **Dimensions** layer current (**Home/Layers** panel).

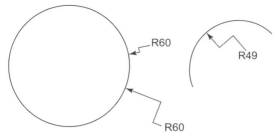

**Fig. 6.16** Examples – the **Jogged** tool.

3. Call the **Jogged** tool, either with a *left-click* on its tool icon in the **Annotate/Dimensions** panel see (Fig. 6.1) or by *entering* **jog** at the command line. The command line shows:

```
Command: _dimjogged
Select arc or circle: pick the circle or the arc
Specify center location override: pick
Dimension text = 60
Specify dimension line location or [Mtext/Text/
   Angle]: pick
Specify jog location: pick
Command:
```

The results of placing a jogged dimension on a circle and an arc are shown in Fig. 6.16.

## DIMENSION TOLERANCES

Before simple tolerances can be included with dimensions, new settings will need to be made in the **Dimension Style Manager** dialog as follows:

1. Open the dialog. The quickest way of doing this is to *enter* **d** at the command line followed by a *right-click*. This opens up the dialog.
2. *Click* the **Modify...** button of the dialog, followed by a *left-click* on the **Primary Units** tab and in the resulting sub-dialog make settings as shown in Fig. 6.17. Note the changes in the preview box of the dialog.

## EXAMPLE – TOLERANCES (FIG. 6.19)

1. Construct the outline Fig. 6.18.
2. Make the **Dimensions** layer current (**Home/Layers** panel).
3. Dimension the drawing using tools from the **Dimensions** panel or by *entering* abbreviations at the command line. Because

**Fig. 6.17** The **Tolerances** sub-dialog of the **Modify Dimension Style** dialog.

**Fig. 6.18** First example – simple tolerances – outline.

tolerances have been set in the **Dimension Style Manager** dialog (Fig. 6.17), the tolerance dimensions will automatically be added to the drawing (Fig. 6.19).

## TEXT

There are two main methods of adding text to drawings – **Multiline Text** and **Single Line Text**.

## EXAMPLE – SINGLE LINE TEXT (FIG. 6.19)

1. Open the drawing from the example on tolerances – Fig. 6.19.
2. Make the **Text** layer current (**Home/Layers** panel).
3. At the command line *enter* **dt** (for Single Line Text) followed by a *right-click*:

```
Command: enter dt right-click
TEXT
```

```
Current text style "ARIAL" Text height: 8 Annotative
  No:
Specify start point of text or [Justify/Style]: pick
Specify rotation angle of text <0>: right-click
Enter text: enter The dimensions in this drawing
  show tolerances press the Return key twice
Command:
```

The result is given in Fig. 6.19.

The dimensions in this drawing show tolerances

**Fig. 6.19** Example – tolerances.

## NOTES →

1. When using **Dynamic Text** the **Return** key of the keyboard is pressed when the text has been *entered*. A *right-click* does not work.

2. At the prompt:

```
Specify start point of text or [Justify/Style]:
  enter s (Style) right-click
Enter style name or [?] <ARIAL>: enter ? right-
  click
Enter text style(s) to list <*>: right-click
```

And an **AutoCAD Text Window** (Fig. 6.20) appears listing all the styles which have been selected in the **Text Style** dialog (see page 82).

3. In order to select the required text style its name must be *entered* at the prompt:

```
Enter style name or [?] <ARIAL>: enter Romand
  right-click
```

And the text *entered* will be in the **Romand** style of height **9**. But only if that style was previously selected in the **Text Style** dialog.

**Fig. 6.20** The **AutoCAD Text Window**.

4. Fig. 6.21 shows some text styles from the **AutoCAD Text Window**.

*This is the TIMES text*

This is ROMANC text

This is ROMAND text

This is STANDARD text

*This is ITALIC text*

This is ARIAL text

**Fig. 6.21** Some text fonts.

5. There are two types of text fonts available in AutoCAD 2012 – the **AutoCAD SHX** fonts and the **Windows True Type** fonts. The **ITALIC, ROMAND, ROMANC,** and **STANDARD** styles shown in Fig. 6.21 are AutoCAD text fonts. The **TIMES** and **ARIAL** styles are **Windows True Type** styles. Most of the **True Type** fonts can be *entered* in **Bold, Bold Italic, Italic,** or **Regular** styles, but these variations are not possible with the AutoCAD fonts.

## → NOTES (CONTINUED)

6. The **Font name** popup list of the **Text Style** dialog shows that a large number of text styles are available to the AutoCAD 2012 operator. It is advisable to practice using a variety of these fonts to familiarize oneself with the text opportunities available with AutoCAD 2012.

## EXAMPLE – MULTILINE TEXT (FIG. 6.23)

1. Make the **Text** layer current (**Home/Layers** panel).
2. Either *left-click* on the **Multiline Text** tool icon in the **Annotate/ Text** panel (Fig. 6.22); or *enter* **t** at the command line:

**Fig. 6.22** Selecting **Multiline Text...** from the **Home/Annotate** panel.

```
Command: _mtext
Current text style: "Arial" Text height: 6
   Annotative No
Specify first corner: pick
Specify opposite corner or [Height/Justify/Line
   spacing/Rotation/Style/Width/Columns]: pick
```

As soon as the **opposite corner** is *picked*, the **Text Formatting** box appears (Fig. 6.23). Text can now be *entered* as required within the box as indicated in this illustration.

**Fig. 6.23** Example – **Multiline Text** *entered* in the text box.

When all the required text has been *entered left-click* and the text box disappears leaving the text on screen.

## SYMBOLS USED IN TEXT

When text has to be added by *entering* letters and figures as part of a dimension, the following symbols must be used:

To obtain Ø75*enter*%%c75;
To obtain 55%*enter*55%%%;
To obtain ±0.05*enter*%%p0.05;
To obtain 90°*enter*90%%d.

## CHECKING SPELLING

### FIRST EXAMPLE – SPELL CHECKING – DDEDIT (FIG. 6.24)

1. *Enter* some badly spelt text as indicated in Fig. 6.24.

Thiss shows somme baddly spelt text

**1. The mis-spelt text**

Thiss shows somme baddly spelt text

**2. Text is selected**

This shows some badly spelt text

**3. The text after correction**

Fig. 6.24 First example – **spell checking – ddedit**.

2. *Enter* **ddedit** at the command line.
3. *Left-click* on the text. The text is highlighted. Edit the text as if working in a word processing application and when satisfied *left-click* followed by a *right-click*.

# SECOND EXAMPLE – THE SPELLING TOOL (FIG. 6.25)

1. *Enter* some badly spelled text as indicated in Fig. 6.25.

**Fig. 6.25** Second example – the **Check Spelling** dialog.

2. *Enter* **spell** or **sp** at the command line.
3. The **Check Spelling** dialog appears (Fig. 6.25). In the **Where to look** field select **Entire drawing** from the field's popup list. The first badly spelled word is highlighted with words to replace them listed in the **Suggestions** field. Select the appropriate correct spelling as shown. Continue until all text is checked. When completely checked an **AutoCAD Message** appears (Fig. 6.26) If satisfied *click* its **OK** button.

**Fig. 6.26** The **AutoCAD Message** window showing that the spelling check is complete.

## REVISION NOTES

1. In the Line and Arrows sub-dialog of the Dimension Style Manager dialog Lineweights were set to 0.3. If these Lineweights are to show in the drawing area of AutoCAD 2012, the Show/Hide Lineweight button in the status bar must be set ON.
2. Dimensions can be added to drawings using the tools from the Annotate/Dimensions panel, or by *entering* **dim**, followed by abbreviations for the tools at the command line.

## REVISION NOTES (CONTINUED)

3. It is usually advisable to use Osnaps when locating points on a drawing for dimensioning.

4. The Style and Angle of the text associated with dimensions can be changed during the dimensioning process.

5. When wishing to add tolerances to dimensions it will probably be necessary to make new settings in the Dimension Style Manager dialog.

6. There are two methods for adding text to a drawing – Single Line Text and Multiline Text.

7. When adding Single Line Text to a drawing, the Return key must be used and not the right-hand mouse button.

8. Text styles can be changed during the process of adding text to drawings.

9. AutoCAD 2012 uses two types of text style – AutoCAD SHX fonts and Windows True Type fonts.

10. Most True Type fonts can be in bold, bold italic, italic, or regular format. AutoCAD fonts can only be added in the single format.

11. To obtain the symbols Ø; ±; °; % use %%c; %%p; %%d;%%% before the figures of the dimension.

12. Text spelling can be checked by selecting Object/Text/Edit... from the Modify drop-down menu, by selecting Spell Check... from the Annotate/Text panel, or by entering spell or sp at the command line.

## EXERCISES

Methods of constructing answers to the following exercises can be found in the free website:

**http://www.routledge.com/cw/yarwood**

1. Open any of the drawings previously saved from working through examples or as answers to exercises and add appropriate dimensions.

2. Construct the drawing in Fig. 6.27 but in place of the given dimensions add dimensions showing tolerances of 0.25 above and below.

**Fig. 6.27** Exercise 2.

3. Construct and dimension the drawing in Fig. 6.28.

**Fig. 6.28** Exercise 3.

4. Construct two polygons as in Fig. 6.29 and add all diagonals. Set Osnaps **endpoint** and **intersection** and, using the lines as in Fig. 6.29, construct the stars as shown using a polyline of width = 3. Next erase all unwanted lines. Dimension the angles labeled **A**, **B,C**, and **D**.

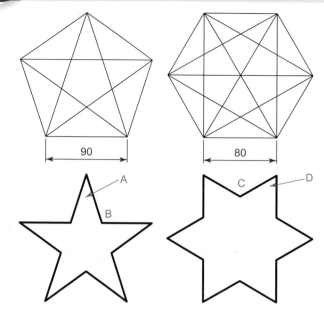

**Fig. 6.29** Exercise 4.

5.  Using the text style **Arial** of height **20** and enclosing the wording within a pline rectangle of Width = 5 and Fillet = 10, construct Fig. 6.30.

AutoCAD 2012

**Fig. 6.30** Exercise 5.

# ORTHOGRAPHIC AND ISOMETRIC

## AIMS OF THIS CHAPTER:

**1.** To describe methods of using layers.
**2.** To explain the value of layers when constructing drawings.
**3.** To introduce methods of constructing views in orthographic projection and the construction of isometric drawings.

## LAYERS

Using layers is an important feature in the successful construction of drawings in AutoCAD. Adding layers to the acadiso.dwt template was described in Chapter 4.

Layers are held in the **Layer Properties Manager**, which, in the acadiso.dwt template used in this book, contains 5 layers (Fig. 7.1). Note in Fig. 7.1 that **Layer 0** is current (green tick preceding its name).

**Fig. 7.1** The **Layer Properties Manager** in the acadiso.dwt template.

**Introduction to AutoCAD 2012. ISBN: 978-0-08-096947-3**

In the **Layer Properties Manager** the following properties for each layer can be set:

**On:** *Click* the on/off icon against a layer name in the **On** list and the layer can be turned on or off (yellow/blue colored icon). If **On**, constructions can be made on the layer. If **Off**, constructions on the layer disappear and the layer cannot be used.

**Freeze:** *Click* the freeze icon against a layer name and the layer is frozen (icon changes shape and becomes blue in color). Freezing a layer makes objects on that layer disappear. Freezing can be used as an alternative to turning a layer off.

**Color:** The color of constructions on a layer take on the color shown against the layer name.

**Linetype:** The linetype of constructions on a current layer.

**Lineweight:** Sets the lineweight of objects on screen constructed on the current layer. Note that the lineweight of objects does not shown on screen unless the **Show/Hide Lineweight** button is set on in the status bar. However, when printed or plotted, objects do print or plot to the lineweight shown against a layer name.

**Transparency:** A figure between **0** and **90** can be set for transparency against a layer name. When a figure higher than 0 is set, objects constructed on the layer show in a transparent form on screen. The larger the number the more transparent the objects become.

**Plot Style:** Shows the color to which the objects on a layer will print or plot.

**Plot:** A *click* on an icon in the **Plot** list causes the red icon to appear against the **Plot** icon, whereby, if a drawing on screen is printed or plotted, objects on that layer will not show in the printout.

**Description:** Any description of a layer can be *entered* in this list.

## The Icons in the Menu Bar of the Dialog

Fig. 7.2 shows the four icons in the menu bar. When wishing to change the status of a layer, first *click* on the required icon, then perform the action – such as making a **New layer**, **Deleting a layer**, or **Making a layer Current**.

## USING THE LAYER PROPERTIES MANAGER

When constructing a drawing on layers, it will be necessary to make current the layer for the linetype being used, its color, or other properties of the layer. There are two main methods of making a selected layer current.

Current layer: 0 : Outlines

**New Layer (Alt+N)**
Creates a new layer. The list displays a layer named LAYER1. The name is selected so that you can enter a new layer name immediately. The new layer inherits the properties of the currently selected layer in the layer list (color, on or off state, and so on).

Current layer: 0 : Outlines

**New Layer VP Frozen in All Viewports**
Creates a new layer and freezes it in all existing layout viewports. This button is accessible from the Model tab or layout tabs.

Current layer: 0 : Outlines

**Delete Layer (Alt+D)**
Deletes selected layers. You can delete only unreferenced layers. Referenced layers include layers 0 and DEFPOINTS, layers containing objects (including objects in block definitions), the current layer, and xref-dependent layers.

Current layer: 0 : Outlines

**Set Current (Alt+C)**
Sets the selected layer as the current layer. Objects that you create are drawn on the current layer. (CLAYER system variable)

**Fig. 7.2** The icons in the menu bar of the **Layer Properties Manager**.

**1.** The **Layer Properties Manager** can be opened with a *click* on its icon in the **Home/Layers** panel (Fig. 7.3) or by *entering* **layer** or **la** at the command line.

Unsaved Layer State

**Layer Properties**
Manages layers and layer properties

**LAYER**
Press F1 for more help

**Fig. 7.3** The **Layer Properties** icon in the **Home/Layers Panel**.

**2.** In the **Home/Layers** panel *click* in the 0 field; when the popup menu showing all the layers appears, *click* again in the field showing the name of the required layer. Fig. 7.4 shows the **Dimensions** layer being made current.

## 7

**Fig. 7.4** Making the **Dimensions** layer current from the **Home/Layers** panel.

## ORTHOGRAPHIC PROJECTION

Orthographic projection involves viewing an article being described in a technical drawing from different directions – from the front, from a side, from above, from below or from any other viewing position. Orthographic projection often involves:

1. The drawing of details are hidden using hidden detail lines.
2. Sectional views in which the article being drawn is imagined as being cut through and the cut surface drawn.
3. Center lines through arcs, circles, spheres, and cylindrical shapes.

## AN EXAMPLE OF AN ORTHOGRAPHIC PROJECTION

**Fig. 7.5** Example: orthographic projection – the solid being drawn.

Taking the solid shown in Fig. 7.5, to construct a three-view orthographic projection of the solid:

1. Draw what is seen when the solid is viewed from its left-hand side and regard this as the **front** of the solid. What is drawn will be a **front view** (Fig. 7.6).
2. Draw what is seen when the solid is viewed from the left-hand end of the front view. This produces an **end view**. Fig. 7.7 shows the end view alongside the front view.
3. Draw what is seen when the solid is viewed from above the front view. This produces a **plan**. Fig. 7.8 shows the plan below the front view.
4. In the **Home/Layers** panel, In the **Layer** list *click* on **Centre** to make it the current layer (Fig. 7.9). All lines will now be drawn as center lines.
5. In the three-view drawing add center lines.

**Fig. 7.6** The solid used to demonstrate **first** and **third angles** of projection (**front** view).

**Fig. 7.7** **Front** and **end** views of the solid.

6. Make the **Hidden** layer the current layer and add hidden detail lines.
7. Make the **Text** layer current and add border lines and a title block.
8. Make the **Dimensions** layer current and add all dimensions.

The completed drawing is shown in Fig. 7.10.

## FIRST ANGLE AND THIRD ANGLE

There are two types of orthographic projection – **first angle** and **third angle**. Fig. 7.11 is a pictorial drawing of the solid used to

**Fig. 7.8** **Front** and **end** views (top) and **plan** of the solid.

**Fig. 7.9** Making the layer **Centre** current from the **Home/Layers** panel.

demonstrate the two angles. Fig. 7.12 shows a three-view **first angle** projection and Fig. 7.13 the same views in **third angle**.

In both angles the viewing is from the same directions. The difference is that the view as seen is placed on the viewing side of the front view in **third angle** and on the opposite side to the viewing in **first angle**.

Dimensions in millimeters

60

110

R20

R10

180

Ø120

Ø60

40

Ø40

30

R40

Hole Ø40

R20

| Name: A. Student | Scale: 1:1 | Date: 19/07/2006 | Title: PART 7/45/EB |
|---|---|---|---|

**Fig. 7.10** The completed working drawing of the solid.

**Fig. 7.11** The solid used to demonstrate **first** and **third angle** projections.

Look from this direction to see a PLAN

Look from this direction to see an END VIEW

Front view

End view

Plan

**Fig. 7.12** A **first angle** projection.

**Fig. 7.13** A **third angle** projection.

## ADDING HATCHING

In order to show internal shapes of a solid being drawn in orthographic projection, the solid is imagined as being cut along a plane and the cut surface is then drawn as seen. Common practice is to **hatch** the areas which then show in the cut surface. Note the section plane line, the section label, and the hatching in the sectional view in Fig. 7.14.

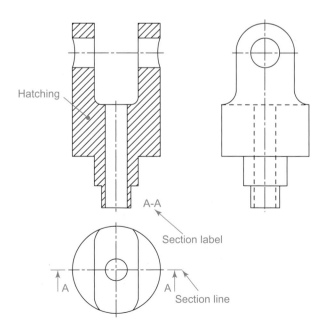

**Fig. 7.14** A sectional view.

To add the hatching as shown in Fig. 7.14:

1. Call the **Hatch** tool with a *left-click* on its tool icon in the **Home/ Draw** panel (Fig. 7.15). A new tab **Hatch Creation** is created and opens the **Hatch Creation** ribbon (Fig. 7.16), but only if the ribbon is active.

**Fig. 7.15** The **Hatch** tool icon and tooltip from the **Home/Draw** panel.

**Fig. 7.16** The **Hatch Creation** tab and ribbon.

2. In the **Hatch Creation/Pattern** panel *click* the bottom arrow on the right of the panel and from the palette which appears *pick* the **ANSI31** pattern (Fig. 7.17).

**Fig. 7.17** Selecting **ANSI31** pattern from the **Hatch Creation/ Pattern** panel.

3. In the **Hatch Creation/Properties** panel, adjust the **Hatch Scale** to **2** (Fig. 7.18).

**Fig. 7.18** Setting the **Hatch Scale** to **2** in the **Hatch Creation/Properties** panel.

4. In the **Hatch Creation/Boundaries** panel *left-click* the **Pick Points** icon (Fig. 7.19).

**Fig. 7.19** Select **Pick Points** from the **Hatch Creation/Boundaries** panel.

5. *Pick* the points in the front view (left-hand drawing of Fig. 7.20) and the *picked* points hatch. If satisfied the hatching is correct *right-click* (right-hand drawing of Fig. 7.20).

**Fig. 7.20** The result of hatching.

## ISOMETRIC DRAWING

### NOTE →

Isometric drawing must not be confused with solid model drawing, examples of which are given in Chapters 12 to 19. Isometric drawing is a 2D method of describing objects in a pictorial form.

## Setting the AutoCAD Window for Isometric Drawing

To set the AutoCAD 2012 window for the construction of isometric drawings:

**1.** At the command line:

```
Command: enter snap
Specify snap spacing or [On/Off/Aspect /Rotate/
   Style/Type] <5>: s (Style)
Enter snap grid style [Standard/Isometric] <S>: i
   (Isometric)
Specify vertical spacing <5>: right-click
Command:
```

And the grid lines in the window assume a pattern as shown in Fig. 7.21. Note also the cursor hair lines which are set at an **Isometric Left** angle.

**Fig. 7.21** The AutoCAD grid points set for isometric drawing.

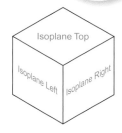

**Fig. 7.22** The three isoplanes.

**Fig. 7.23** The three isocircles.

2. There are three isometric angles – **Isoplane Top, Isoplane Left,** and **Isoplane Right.** These can be set either by pressing the **F5** function key or by pressing the **Ctrl** and **E** keys. Repeated pressing of either of these "toggles" between the three settings. Fig. 7.22 is an isometric view showing the three isometric planes.

### The Isometric Circle

Circles in an isometric drawing show as ellipses. To add an isometric circle to an isometric drawing, call the **Ellipse** tool. The command line shows:

```
Command: _ellipse
Specify axis endpoint of ellipse or [Arc/Center/
   Isocircle]: enter i (Isocircle) right-click
Specify center of isocircle: pick or enter
   coordinates
Specify radius of isocircle or [Diameter]: enter a
   number
Command:
```

And the isocircle appears. Its isoplane position is determined by which of the isoplanes is in operation at the time the isocircle was formed. Fig. 7.23 shows these three isoplanes containing isocircles.

## EXAMPLES OF ISOMETRIC DRAWINGS

### FIRST EXAMPLE – ISOMETRIC DRAWING (FIG. 7.26)

1. This example is to construct an isometric drawing to the details given in the orthographic projection in Fig. 7.24. Set **Snap** on (press the **F9** function key) and **Grid** on (**F7**).
2. Set **Snap** to Isometric and set the isoplane to **Isoplane Top** using **F5**.
3. With **Line**, construct the outline of the top of the model (Fig. 7.25) working to the dimensions given in Fig. 7.24.
4. Call **Ellipse** tool and set to **isocircle** and add the isocircle of radius 20 centered in its correct position in the outline of the top (Fig. 7.25).
5. Set the isoplane to **Isoplane Right** and with the **Copy** tool, copy the top with its ellipse vertically downward three times as shown in Fig. 7.25.
6. Add lines as shown in Fig. 7.25.
7. Finally using **Trim** remove unwanted parts of lines and ellipses to produce Fig. 7.26.

**Fig. 7.24** First example – isometric drawing – the model.

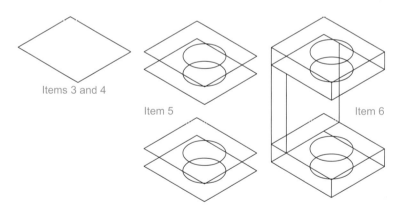

Items 3 and 4

Item 5

Item 6

**Fig. 7.25** First example – isometric drawing – items **3**, **4**, **5**, and **6.**

**Fig. 7.26** First example – isometric drawing.

# SECOND EXAMPLE – ISOMETRIC DRAWING (FIG. 7.28)

Fig. 7.27 shows an orthographic projection of the model from which the isometric drawing is to be constructed.

Fig. 7.28 shows the stages in its construction. The numbers refer to the items in the list below.

**Fig. 7.27** Second example – isometric drawing – orthographic projection.

**Fig. 7.28** Second example – isometric drawing – stages in the construction.

1. In **Isoplane Right** construct two isocircles of radii 10 and 20.
2. Add lines as in drawing **2** and trim unwanted parts of isocircle.
3. With **Copy** copy 3 times as in drawing **3**.
4. With **Trim** trim unwanted lines and parts of isocircle (drawing **4**).
5. In **Isoplane Left** add lines as in drawing **5**.
6. In **Isoplane Right** add lines and isocircles as in drawing **6**.
7. With **Trim** trim unwanted lines and parts of isocircles to complete the isometric drawing – drawing **7**.

## REVISION NOTES

1. There are, in the main, two types of orthographic projection – first angle and third angle.

2. The number of views included in an orthographic projection depends upon the complexity of the component being drawn – a good rule to follow is to attempt fully describing the object in as few views as possible.

3. Sectional views allow parts of an object which are normally hidden from view to be more fully described in a projection.

4. When a layer is turned OFF, all constructions on that layer disappear from the screen.

5. Frozen layers cannot be selected, but note that layer 0 cannot be frozen.

6. Isometric drawing is a 2D pictorial method of producing illustrations showing objects. It is not a 3D method of showing a pictorial view.

7. When drawing ellipses in an isometric drawing the Isocircle prompt of the Ellipse tool command line sequence must be used.

8. When constructing an isometric drawing Snap must be set to Isometric mode before construction can commence.

## EXERCISES

Methods of constructing answers to the following exercises can be found in the free website:

**http://www.routledge.com/cw/yarwood**

**Fig. 7.29** Exercises 1, 2, and 3 – an isometric drawing of the two parts of the slider on which these exercises are based.

**1.** Fig. 7.30 shows a first angle orthographic projection of part of the fitment shown in the isometric drawing in Fig. 7.29. Construct a three-view third angle orthographic projection of the part.

**Fig. 7.30** Exercise 1.

**2.** Fig. 7.31 shows a first angle orthographic projection of the other part of the fitment. Construct a three-view third angle orthographic projection of the part.

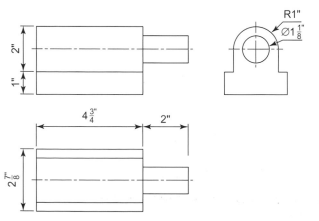

**Fig. 7.31** Exercises 2 and 3.

**3.** Construct an isometric drawing of the part shown in Fig. 7.31.

**4.** Construct a three-view orthographic projection in an angle of your own choice of the tool holder assembled as shown in the isometric drawing in Fig. 7.32. Details are given in Fig. 7.33.

**Fig. 7.32** Exercises 4 and 5 – isometric drawing of the tool holder on which the two exercises are based.

**Fig. 7.33** Exercises 4 and 5 – orthographic projection of the tool holder shown in Fig. 7.32.

**5.** Construct an isometric drawing of the body of the tool holder shown in Fig. 7.32.

**6.** Construct the orthographic projection given in Fig. 7.33.

**7.** Construct an isometric drawing of the angle plate relating to Figs. 7.34 and 7.35.

**Fig. 7.34** An isometric drawing of the angle plate on which exercises 6 and 7 are based.

**Fig. 7.35** Exercises 6 and 7 – an orthographic projection of the angle plate.

**8.** Construct a third angle projection of the component shown in the isometric drawing in Fig. 7.36 and the three-view first angle projection in Fig. 7.37.

**Fig. 7.36** Exercises 8 and 9.

**Fig. 7.37** Exercises 8 and 9.

**9.** Construct the isometric drawing shown in Fig. 7.36 working to the dimensions given in Fig. 7.37.

CHAPTER **8**

# HATCHING

To give further examples of the use of hatching in its various forms.

## INTRODUCTION

In Chapter 7 an example of hatching of a sectional view in an orthographic projection was given. Further examples of hatching will be described in this chapter.

There are a large number of hatch patterns available when hatching drawings in AutoCAD 2012. Some examples from hatch patterns are shown in Fig. 8.1.

**Fig. 8.1** Some hatch patterns from AutoCAD 2012.

Introduction to AutoCAD 2012. ISBN: 978-0-08-096947-3

Other hatch patterns can be selected from **Hatch Creation/Pattern** panel, or the operator can design his/her own hatch patterns as **User Defined** patterns (Fig. 8.2).

**Fig. 8.2** The **User Defined** patterns in the **Hatch Creation/Properties** panel.

## FIRST EXAMPLE – HATCHING A SECTIONAL VIEW (FIG. 8.3)

Fig. 8.3 shows a two-view orthographic projection which includes a sectional end view.

**Fig. 8.3** First example – **Hatching**.

Note the following in the drawing:

1. The section plane line, consisting of a center line with its ends marked **A** and arrows showing the direction of viewing to obtain the sectional view.

2. The sectional view labelled with the letters of the section plane line.
3. The cut surfaces of the sectional view hatched with the **ANSI31** hatch pattern, which is in general use for the hatching of engineering drawing sections.

## SECOND EXAMPLE – HATCHING RULES (FIG. 8.4)

Fig. 8.4 describes the stages in hatching a sectional end view of a lathe tool holder.

**Fig. 8.4** Second example – hatching rules for sections.

Note the following in the section:

1. There are two angles of hatching to differentiate the separate parts of the section.
2. The section follows the general rule that parts such as screws, bolts, nuts, rivets, other cylindrical objects, webs and ribs, and other such features, are shown as outside views within sections.

In order to hatch this example:

1. *Left-click* on the **Hatch** tool icon in the **Home/Draw** panel (Fig. 8.5). The ribbon changes to the **Hatch Creation** ribbon. *Entering* **hatch** or **h** at the command line has the same result.

8

Fig. 8.5 *Left-click* on the **Hatch** tool icon in the **Home/Draw** panel.

2. *Left-click* **ANSI31** in the **Hatch Creation/Pattern** panel (Fig. 8.6).

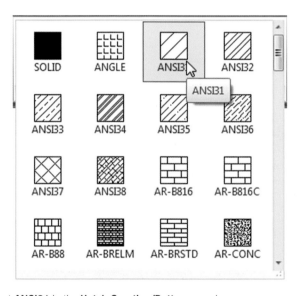

Fig. 8.6 Select **ANSI31** in the **Hatch Creation/Pattern** panel.

3. Set the **Hatch Scale** to **1.5** in the **Hatch Creation/Properties** panel (Fig. 8.7).

Fig. 8.7 Set the **Hatch Scale** in the **Hatch Creation/Properties** panel.

**4.** *Left-click* **Pick Points** in the **Hatch Creation/Boundaries** panel and *pick* inside the areas to be hatched.

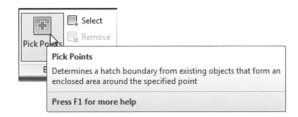

Fig. 8.8 *Left-click* **Pick Points** in the **Hatch Creation/ Boundaries** panel.

**5.** The *picked* areas hatch. If satisfied with the hatching *right-click*. If not satisfied amend the settings and when satisfied *right-click*.

## THE HATCH AND GRADIENT DIALOG

If the ribbon is not on screen, e.g. as when working in the **AutoCAD Classic** workspace, *entering* **hatch** or **h** at the command line brings the **Hatch and Gradient** dialog to screen (Fig. 8.9). The method of hatching given in the previous two examples is much the same whether using the tools in the **Hatch Creation** ribbon or using the **Hatch and Gradient** dialog. Fig. 8.9 shows the **ANSI Hatch Pattern** dialog and the **Pick Points** button in the **Hatch and Gradient** dialog, which are *picked* for the same methods as described in the given examples.

Fig. 8.9 The **Hatch and Gradient** dialog.

## THIRD EXAMPLE – ASSOCIATIVE HATCHING (FIG. 8.10)

Fig. 8.10 shows a two end view of a house. After constructing the left-hand view, it was found that the upper window had been placed in the wrong position.

Fig. **8.10** Third example – **Associative hatching**.

Using the **Move** tool, the window was moved to a new position. The brick hatching automatically adjusted to the new position. Such **associative hatching** is only possible if check box is **ON** – a tick in the check box in the **Options** area of the **Hatch and Gradient** dialog (Fig. 8.11).

Fig. **8.11** **Associative Hatching** set **ON** in the **Hatch and Gradient** dialog.

## FOURTH EXAMPLE – COLOR GRADIENT HATCHING (FIG. 8.12)

Fig. 8.12 shows two examples of hatching from the **Gradient** sub-dialog of the **Hatch and Gradient** dialog.

1. Construct two outlines each consisting of six rectangles (Fig. 8.12).
2. *Click* **Gradient** in the drop-down menu in the **Hatch Creation/Properties** panel (Fig. 8.13).

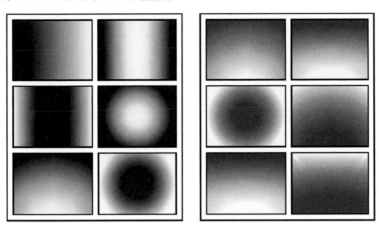

**Fig. 8.12** Fourth example – Color gradient hatching.

**Fig. 8.13** Selecting **Gradient** in the **Hatch Creation/Properties** panel.

In the **Hatch Creation/Pattern** panel which then appears, *pick* one of the gradient choices (Fig. 8.14), followed by a *click* in a single area of one of the rectangles in the left-hand drawing, followed by a *right-click*.

**Fig. 8.14** The **Gradient patterns** in the **Hatch Creation/Pattern** panel.

3. Repeat in each of the other rectangles of the left-hand drawing changing the pattern in each of the rectangles.
4. Change the color of the **Gradient** patterns with a *click* on the red option in the **Select Colors...** drop-down menu in the **Hatch Creation/Properties** panel. The hatch patterns all change color to red (Fig. 8.15).

**Fig. 8.15** Changing the colors of the **Gradient** patterns.

## FIFTH EXAMPLE – ADVANCED HATCHING (FIG. 8.17)

*Left-click* **Normal Island Detection** in the **Hatch Creation/Options** panel extension. The drop-down shows several forms of **Island** hatching (Fig. 8.16).

1. Construct a drawing which includes three outlines as shown in the left-hand drawing of Fig. 8.17 and copy it twice to produce three identical drawings.
2. Select the hatch patterns **STARS** at an angle of **0** and scale **1**.
3. *Click* **Normal Island Detection** from the drop-down menu.
4. *Pick* a point in the left-hand drawing. The drawing hatches as shown.
5. Repeat in the center drawing with **Outer Island Detection** selected.
6. Repeat in the right-hand drawing with **Ignore Island Detection** selected.

**Fig. 8.16** The **Island detection** options in the **Hatch Creation/Options** panel.

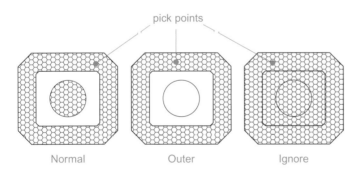

pick points

Normal      Outer      Ignore

**Fig. 8.17** Fifth example – advanced hatching.

## SIXTH EXAMPLE – TEXT IN HATCHING (FIG. 8.18)

1. Construct a pline rectangle using the sizes given in Fig. 8.18.
2. In the **Text Style Manager** dialog, set the text font to **Arial** and its **Height = 25**.
3. Using the **Dtext** tool *enter* the text as shown central to the rectangle.
4. Hatch the area using the **HONEY** hatch pattern set to an angle of **0** and scale of **1**.

The result is shown in Fig. 8.18.

**Fig. 8.18** Sixth example – Text in hatching.

---

NOTE ➡

Text will be entered with a surrounding boundary area free from hatching provided **Normal Island Detection** has been selected from the **Hatch Creation/Options** panel.

---

REVISION NOTES ↻

1. A large variety of hatch patterns are available when working with AutoCAD 2012.
2. In sectional views in engineering drawings it is usual to show items such as bolts, screws, other cylindrical objects, webs, and ribs as outside views.
3. When Associative hatching is set on, if an object is moved within a hatched area, the hatching accommodates to fit around the moved object.
4. Color gradient hatching is available in AutoCAD 2012.
5. When hatching takes place around text, a space around the text will be free from hatching.

## EXERCISES

Methods of constructing answers to the following exercises can be found in the free

website: **http://www.routledge.com/cw/yarwood**

**1.** Fig. 8.19 is a pictorial drawing of the component shown in the orthographic projection in Fig. 8.20. Construct the three views but with the front view as a sectional view based on the section plane **A-A**.

**Fig. 8.19** Exercise 1 – a pictorial view.

SECTION A-A

| M. R. AUTOCAD | Scale 1:1 | 15:06:2007 | ANSWER to CHAPTER 8 EXERCISE 1 |

**Fig. 8.20** Exercise 1.

**2.** Construct the orthographic projection in Fig. 8.21 to the given dimensions with the front view as the sectional view **A-A**.

**3.** Construct the drawing **Stage 5** in Fig. 8.22 following the descriptions of stages given.

**Fig. 8.21** Exercise 2.

Stage 1
Construct word
on Layer 0 and
offset on Layer 1

Stage 2
Hatch on Layer
HATCH01
with SOLID
Turn Layer 0 off

Stage 3
Turn HATCH01 off
Turn Layer
HATCH02 on
Add lines as shown

Stage 4
On HATCH03
Hatch with ANSI31
at Angle 135 and
Scale 40
Turn HATCH02 off

Stage 5
Turn HATCH02 on

**Fig. 8.22** Exercise 3.

**4.** Fig. 8.23 shows a side view of a car with parts hatched. Construct a similar drawing of any make of car, using hatching to emphasize the shape.

**5.** Working to the notes given with the drawing in Fig. 8.24, construct the end view of a house as shown. Use your own discretion about sizes for the parts of the drawing.

**Fig. 8.23** Exercise 4.

**Fig. 8.24** Exercise 5.

**6.** Working to dimensions of your own choice, construct the three-view projection of a two-story house as shown in Fig. 8.25.

**7.** Construct Fig. 8.26 as follows:

**(a)** On layer **Text**, construct a circle of radius **90**.

**(b)** Make layer **0** current.

**(c)** Construct the small drawing to the details as shown and save as a block with a block name **shape** (see Chapter 9).

**(d)** Call the **Divide** tool by *entering* **div** at the command line:

```
Command: enter div right-click
Select object to divide: pick the circle
Enter number of segments or [Block]: enter b right-click
Enter name of block to insert: enter shape right-click
Align block with object? [Yes/No] <Y>: right-click
Enter the number of segments: enter 20 right-click
Command
```

**(e)** Turn the layer **Text** off.

**Fig. 8.25** Exercise 6.

**Fig. 8.26** Exercise 7.

CHAPTER **9**

# BLOCKS AND INSERTS

## AIMS OF THIS CHAPTER

The aims of this chapter are:

**1.** To describe the construction of **blocks** and **wblocks** (written blocks).
**2.** To introduce the insertion of blocks and wblocks into drawings.
**3.** To introduce uses of the **DesignCenter** palette.
**4.** To explain the use of the **Explode** and **Purge** tools.

## INTRODUCTION

Blocks are drawings which can be inserted into other drawings. Blocks are contained in the data of the drawing in which they have been constructed. Wblocks (written blocks) are saved as drawings in their own right, but can be inserted into other drawings if required.

## BLOCKS

### FIRST EXAMPLE – BLOCKS (FIG. 9.3)

**1.** Construct the building symbols as shown in Fig. 9.1 to a scale of 1:50.
**2.** *Left-click* the **Create** tool icon in the **Home/Block** panel (Fig. 9.2).
The **Block Definition** dialog (Fig. 9.3) appears. To make a block from the **Compass** symbol drawing:
**(a)** *Enter* **compass** in the **Name** field.

**Fig. 9.1** First example – **Blocks** – symbols to be saved as blocks.

**Fig. 9.2** *Left-click* **Create** tool icon in the **Home/Block** panel.

**(b)** *Click* the **Select Objects** button. The dialog disappears. Window the drawing of the compass. The dialog reappears. Note the icon of the compass at the top-center of the dialog.

**(c)** *Click* the **Pick Point** button. The dialog disappears. *Click* a point on the compass drawing to determine its **insertion point**. The dialog reappears.

**Fig. 9.3** The **Block Definition** dialog with *entries* for the **compass** block.

**(d)** If thought necessary *enter* a description in the **Description** field of the dialog.

**(e)** *Click* the **OK** button. The drawing is now saved as a **block** in the drawing.

3. Repeat items **1** and **2** to make blocks of all the other symbols in the drawing.

4. Open the **Block Definition** dialog again and *click* the arrow on the right of the **Name** field. Blocks saved in the drawing are listed (Fig. 9.4).

**Fig. 9.4** The popup list in the **Name** field of the **Block Definition** dialog.

## INSERTING BLOCKS INTO A DRAWING

There are two methods by which symbols saved as blocks, can be inserted into another drawing.

## EXAMPLE – FIRST METHOD OF INSERTING BLOCKS

Ensure that all the symbols saved as blocks using the **Create** tool are saved in the data of the drawing in which the symbols were constructed. Erase all of the drawings of the symbols and in their place construct the outline of the plan of a bungalow to a scale of 1:50 (Fig. 9.5).

Then:

1. *Left-click* the **Insert** tool icon in the **Home/Block** panel (Fig. 9.6) or the **Insert Block** tool in the **Draw** toolbar.

**Fig. 9.5** First example – inserting blocks. Outline plan.

**Fig. 9.6** The **Insert** tool icon in the **Home/Block** panel.

The **Insert** dialog appears on screen (Fig. 9.7). From the **Name** popup list select the name of the block which is to be inserted, in this example the **2.5 window**.

**Fig. 9.7** The **Insert** dialog with its **Name** pop-up list showing all the blocks.

**2.** *Click* the dialog's **OK** button; the dialog disappears. The symbol drawing appears on screen with its insertion point at the intersection of the cursor hairs ready to be *dragged* into its position in the plan drawing.

3. Once all the block drawings are placed, their positions can be adjusted. Blocks are single objects and can thus be dragged into new positions as required under mouse control. Their angle of position can be amended at the command line, which shows:

```
Command: _insert
Specify insertion point or [Basepoint/Scale/Rotate]:
  pick
Command:
```

Selection from these prompts allows scaling or rotating as the block is inserted.

4. Insert all necessary blocks and add other detail as required to the plan outline drawing. The result is given in Fig. 9.8.

**Fig. 9.8** Example – first method of inserting blocks.

## EXAMPLE – SECOND METHOD OF INSERTING BLOCKS

1. Save the drawing with all the blocks to a suitable file name. Remember this drawing includes data of the blocks in its file.
2. *Left-click* **DesignCenter** in the **View/Palettes** panel (Fig. 9.9) or press the **Ctrl+2** keys. The **DesignCenter** palette appears on screen (Fig. 9.10). **Center** with the compass block *dragged* on screen.
3. With the outline plan (Fig. 9.5) on screen the symbols can all be *dragged* into position from the **DesignCenter**.

**Fig. 9.9** Selecting **DesignCenter** from the **View/Palettes** panel.

**Fig. 9.10** The **DesignCenter** with the compass block *dragged* on screen.

## NOTES ABOUT THE DESIGNCENTER PALETTE →

1. As with other palettes, the **DesignCenter** palette can be re-sized by *dragging* the palette to a new size from its edges or corners.

2. *Clicks* on one of the three icons at the top-right corner of the palette (Fig. 9.11) have the following results:

**Fig. 9.11** The icons at the top of the **DesignCenter** palette.

**Tree View Toggle:** changes from showing two areas, a **Folder List** and icons of the blocks within a file and icons of the blocks within a file, to a single area showing the block icons (Fig. 9.12).

## → NOTES (CONTINUED)

**Fig. 9.12** The results of a *click* on **Tree View Toggle**.

**Preview:** a *click* on the icon opens a small area at the base of the palette open showing an enlarged view of the selected block icon.

**Description:** a *click* on the icon opens another small area with a description of the block.

A block is a single object no matter from how many objects it was originally constructed. This enables a block to be *dragged* about the drawing area as a single object.

## THE EXPLODE TOOL

**Fig. 9.13** The **Explode** check box in the **Insert** dialog.

A check box in the bottom left-hand corner of the **Insert** dialog is labeled **Explode**. If a tick is in the check box, **Explode** will be set on and when a block is inserted it will be exploded into the objects from which it was constructed (Fig. 9.13).

Another way of exploding a block would be to use the **Explode** tool from the **Home/Modify** panel (Fig. 9.14).

A *click* on the icon or *entering* **ex** at the command line brings prompts into the command line:

```
Command: _explode
Select objects: <Object Snap Tracking on> pick a
 block on screen 1 found
Select objects: right-click
Command:
```

And the *picked* object is exploded into its original objects.

**Fig. 9.14** The **Explode** tool icon in the **Home/Modify** panel.

## PURGE

The **Purge** dialog (Fig. 9.15) is called to screen by *entering* **pu** or **purge** at the command line.

**Fig. 9.15** The **Purge** dialog.

**Purge** can be used to remove data (if any is to be purged) from within a drawing, thus saving file space when a drawing is saved to disk.

To purge a drawing of unwanted data (if any), in the dialog *click* the **Purge All** button and a sub-dialog appears with three suggestions: purging of a named item, purging of all the items, or skip purging a named item.

Take the drawing in Fig. 9.8 (page 179) as an example. If all the unnecessary data is purged from the drawing, the file will be reduced from **145** kbytes to **67** kbytes when the drawing is saved to disk.

## USING THE DESIGNCENTER (FIG. 9.18)

1.  Construct the set of electric/electronic circuit symbols shown in Fig. 9.16 and make a series of blocks from each of the symbols.

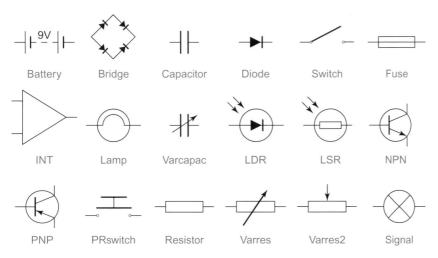

**Fig. 9.16** Example using the DesignCenter – electric/electronic symbols.

2.  Save the drawing to a file designated Fig16.dwg.
3.  Open the **acadiso.dwt** template. Open the **DesignCenter** with a *click* on its icon in the **View/Palettes** panel.
4.  From the **Folder list** select the file Fig16.dwg and *click* on **Blocks** under its file name. Then *drag* symbol icons from the **DesignCenter** into the drawing area as shown in Fig. 9.17. Ensure they are placed in appropriate positions in relation to

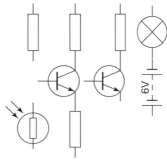

**Fig. 9.17** Example using the **DesignCenter**.

each other to form a circuit. If necessary either **Move** or/and **Rotate** the symbols into correct positions.

5. Close the **DesignCenter** palette with a *click* on the **x** in the top left-hand corner.

6. Complete the circuit drawing as shown in Fig. 9.18.

**Fig. 9.18** Example using the DesignCenter.

It should be noted here that Fig. 9.18 does not represent an authentic electronics circuit.

## WBLOCKS

Wblocks, or written blocks, are saved as drawing files in their own right and are not part of the drawing in which they have been saved.

## EXAMPLE – WBLOCK (FIG. 9.19)

1. Construct a light emitting diode (**LED**) symbol and *enter* **w** at the command line. The **Write Block** dialog appears (Fig. 9.19).

2. *Click* the button marked with three full points (**...**) to the right of the **File name and path** field, and from the **Browse for Drawing File** dialog which comes to screen select an appropriate directory. The directory name appears in the **File name and path** field. Add **LED.dwg** at the end of the name.

3. Make sure the **Insert units** is set to **Millimeters** in its popup list.

4. *Click* the **Select objects** button, Window the symbol drawing and when the dialog reappears, *click* the **Pick point** button, followed by selecting the left-hand end of the symbol.

5. Finally *click* the **OK** button of the dialog and the symbol is saved in its selected directory as a drawing file **LED.dwg** in its own right.

**Fig. 9.19** Example – **Wblock**.

## NOTE ON THE DESIGNCENTER →

Drawings can be inserted into the AutoCAD window from the **DesignCenter** by *dragging* the icon representing the drawing into the window (Fig. 9.20).

When such a drawing is *dragged* into the AutoCAD window, the command line shows a sequence such as:

**Fig. 9.20** An example of a drawing *dragged* from the **DesignCenter**.

## ➔ NOTE (CONTINUED)

```
Command: _-INSERT Enter block name or [?]: "C:
  \Acad 2012 book\Chapter11\64
Pheasant Drive\Fig04.dwg"
Units: Millimeters Conversion: 1.0000
Specify insertion point or [Basepoint/Scale/X/Y/Z/
  Rotate]: pick
Enter X scale factor, specify opposite corner, or
  [Corner/XYZ] <1>: right-click
Enter Y scale factor <use X scale factor>: right-
  click
Specify rotation angle <0>: right-click
Command:
```

## REVISION NOTES ↻

1. Blocks become part of the drawing file in which they were constructed.
2. Wblocks become drawing files in their own right.
3. Drawings or parts of drawings can be inserted in other drawings with the Insert tool.
4. Inserted blocks or drawings are single objects unless either the Explode check box of the Insert dialog is checked or the block or drawing is exploded with the Explode tool.
5. Drawings can be inserted into the AutoCAD drawing area using the DesignCenter.
6. Blocks within drawings can be inserted into drawings from the DesignCenter.
7. Construct drawings of the electric/electronics symbols in Fig. 9.16 and save them as blocks in a drawing file designated electronics.dwg.

## EXERCISES

Methods of constructing answers to the following exercises can be found in the free website:

**http://www.routledge.com/cw/yarwood**

1.  Construct the building symbols in Fig. 9.21 in a drawing saved as **symbols.dwg**. Then using the **DesignCenter** construct a building drawing of the first floor of the house you are living in making use of the symbols. Do not bother too much about dimensions because this exercise is designed to practice using the idea of making blocks and using the **DesignCenter**.

**Fig. 9.21** Exercise 1.

2.  Construct drawings of the electric/electronics symbols in Fig. 9.16 (page 183) and save them in a drawing file **electronics.dwg**.

3.  Construct the electronics circuit given in Fig. 9.22 from the file **electronics.dwg** using the **DesignCenter**.

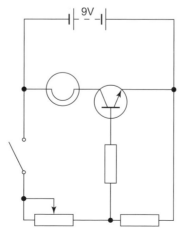

**Fig. 9.22** Exercise 3.

4.   Construct the electronics circuit given in Fig. 9.23 from the file **electronics.dwg** using the **DesignCenter**.

**Fig. 9.23** Exercise 4.

# CHAPTER 10

# OTHER TYPES OF FILE FORMAT

## AIMS OF THIS CHAPTER

The aims of this chapter are:

1. To introduce Object Linking and Embedding (**OLE**) and its uses.
2. To introduce the use of Encapsulated Postscript (**EPS**) files.
3. To introduce the use of Data Exchange Format (**DXF**) files.
4. To introduce raster files.
5. To introduce **Xrefs**.

## OBJECT LINKING AND EMBEDDING

### FIRST EXAMPLE – COPYING AND PASTING (FIG. 10.2)

1. Open any drawing in the AutoCAD 2012 window (Fig. 10.1).
2. *Click* **Copy Clip** from the **Home/Clipboard** panel. The command line shows:

```
Command: _copyclip
Select objects: left-click top left of the drawing
Specify opposite corner: left-click bottom right of the
  drawing 457 found
Select objects: right click
Command:
```

3. Open **Microsoft Word** and *click* on **Paste** in the **Edit** drop-down menu (Fig. 10.2). The drawing from the **Clipboard** appears in the **Microsoft Word** document. Add text as required.

---

**Introduction to AutoCAD 2012. ISBN: 978-0-08-096947-3**

**Fig. 10.1** A drawing in the AutoCAD 2012 with **Copy Clip** selected.

**Fig. 10.2** Example – **Copying and Pasting.**

NOTE →

Similar results can be obtained using the **Copy, Copy Link,** or **Copy with Base Point** tools from the **Edit** drop-down menu.

# SECOND EXAMPLE – eps FILE (FIG. 10.5)

1.  With the same drawing on screen *click* on **Export...** in the **File** drop-down menu (Fig. 10.3) or *click* **Export/Other formats** in the menu appearing with a *click* on the **A** icon at the top-left-hand corner of the AutoCAD window. The **Export Data** dialog appears (Fig. 10.3). *Pick* **Encapsulated PS (\*.eps)** from the **Files of type** popup list then *enter* a suitable file name (e.g. **building.eps**) in the **File name field** and *click* the **Save** button.

**Fig. 10.3** The **Export** tool icon from the **File** drop-down menu and the **Export Data** dialog.

2.  Open a desktop publishing application. That shown in Fig. 10.4 is **PageMaker**.
3.  From the **File** drop-down menu of **PageMaker** *click* **Place...** . A dialog appears listing files which can be placed in a **PageMaker** document. Among the files named will be **building.eps**. *Double-click* that file name and an icon appears the placing of which determines the position of the **\*.eps** file drawing in the **PageMaker** document (Fig. 10.4).
4.  Add text as required.
5.  Save the PageMaker document to a suitable file name.
6.  Go back to the AutoCAD drawing and delete the title.
7.  Make a new **\*.eps** file with the same file name (**building.eps**).
8.  Go back into **PageMaker** and *click* **Links Manager...** in the drop-down menu. The **Links Manager** dialog appears (Fig. 10.5). Against the name of the **building.eps** file name is a dash and a note at the bottom of the dialog explaining that changes have taken place in the drawing from which the **\*.eps** had been derived. *Click the* **Update** button and when the document reappears the drawing in **PageMaker** no longer includes the erased title.

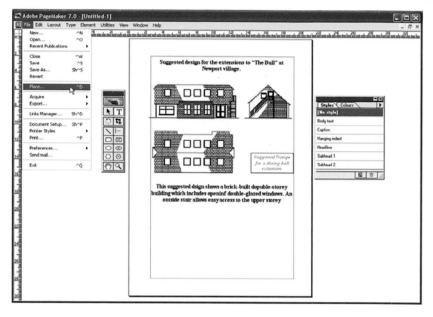

**Fig. 10.4** An *.eps file placed in position in a **PageMaker** document.

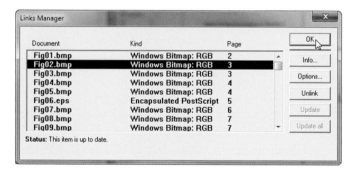

**Fig. 10.5** The **Links Manager** dialog of **PageMaker**.

## NOTES →

1. This is **Object Linking and Embedding** (OLE). Changes in the AutoCAD drawing saved as an *.eps file are linked to the drawing embedded in another application document, so changes made in the AutoCAD drawing are reflected in the PageMaker document.

2. There is actually no need to use the **Links Manager** because if the file from PageMaker is saved with the old *.eps file in place, when it is reopened the file will have changed to the redrawn AutoCAD drawing, without the erased title.

## DXF (DATA EXCHANGE FORMAT) FILES

The *.**DXF** format was originated by Autodesk (publishers of AutoCAD), but is now in general use in most **CAD** (Computer Aided Design) software. A drawing saved to a *.**dxf** format file can be opened in most other CAD software applications. This file format is of great value when drawings are being exchanged between operators using different CAD applications.

## EXAMPLE – DXF FILE (FIG. 10.7)

**1.** Open a drawing in AutoCAD. This example is shown in Fig. 10.6.

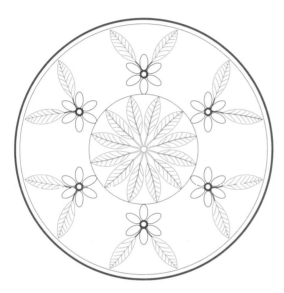

**Fig. 10.6** Example – **DXF file**. Drawing to be saved as a .dxf file.

**2.** *Click* on **Save As...** in the **Menu Browser** dialog and in the **Save Drawing As** dialog which appears, *click* **AutoCAD 2010 DXF [*.dxf]** in the **Files of type** field popup list.
**3.** *Enter* a suitable file name. In this example this is **Fig06.dxf**. The extension **.dxf** is automatically included when the **Save** button of the dialog is *clicked* (Fig. 10.7).
**4.** The **DXF** file can now be opened in the majority of CAD applications and then saved to the drawing file format of the CAD in use.

## NOTE →

To open a **DXF** file in AutoCAD 2012, select **Open...** from the **Menu Browser** dialog and in the **Select File** dialog select **DXF [*.dxf]** from the popup list from the **Files of type** field.

**Fig. 10.7** The **Save Drawing As** dialog set to save drawings in **DXF** format.

**Fig. 10.8** Selecting **Raster Image Reference...** from the **Insert** drop-down menu.

## RASTER IMAGES

A variety of raster files can be placed into AutoCAD 2012 drawings from the **Select Image File** dialog brought to screen with a *click* on **Raster Image Reference...** from the **Insert** drop-down menu. In this example the selected raster file is a bitmap (extension **\*.bmp**) of a rendered 3D model drawing.

## EXAMPLE – PLACING A RASTER FILE IN A DRAWING (FIG. 10.11)

1. *Click* **Raster Image Reference...** from the **Insert** drop-down menu (Fig. 10.8). The **Select Reference File** dialog appears (Fig. 10.9). *Click* the file name of the image to be inserted, **Fig05** (a bitmap \*.bmp). A preview of the bitmap appears.
2. *Click* the **Open** button of the dialog. The **Attach Image** dialog appears (Fig. 10.10) showing a preview of the bitmap image.
3. *Click* the **OK** button; the command line shows:

```
Command: _imageattach
Specify insertion point <0,0>: click at a point on
    screen
Base image size: Width: 1.000000, Height: 1.032895,
    Millimeters
Specify scale factor <1>: drag a corner of the image
    to obtain its required size
Command:
```

Fig. 10.9 The **Select Reference File** dialog.

Fig. 10.10 The **Attach Image** dialog.

And the raster image appears at the *picked* point (Fig. 10.11).

Fig. 10.11 Example – placing a raster file in a drawing.

## NOTES →

As will be seen from the **Insert** drop-down menu and the dialogs which can be opened from the menu, a variety of different types of images can be inserted into an AutoCAD drawing. Some examples are:

**External References** (Xrefs): If a drawing is inserted into another drawing as an external reference, any changes made in the original Xref drawing are automatically reflected in the drawing into which the Xref has been inserted. See later in this chapter.

**Field:** A *click* on the name brings up the **Field** dialog. Practice inserting various categories of field names from the dialog.

**Layout:** The **Field** dialog appears allowing new text to be created and inserted into a drawing.

**3D Studio:** allows the insertion of images constructed in the Autodesk software **3D Studio** from files with the format *.3ds.

## EXTERNAL REFERENCES (XREFS)

If a drawing is inserted into another drawing as an external reference, any changes made in the original Xref drawing subsequent to its being inserted, are automatically reflected in the drawing into which the Xref has been inserted.

## EXAMPLE – EXTERNAL REFERENCES (FIG. 10.19)

1. Construct the three-view orthographic drawing in Fig. 10.12. Dimensions for this drawing will be found in Fig. 15.56 (page 316). Save the drawing to a suitable file name.

2. As a separate drawing construct Fig. 10.13. Save it as a wblock with the name of **Fig13.dwg** and with a base insertion point at the crossing of its center line with the left-hand end of its spindle.

3. *Click* **External References** in the **View/Palettes** panel (Fig. 10.14). The **External References** palette appears (Fig. 10.15).

4. *Click* its **Attach** button and select **Attach DWG...** from the popup list which appears when a *left-click* is held on the button. Select the drawing of a spindle (**Fig13.dwg**) from the **Select Reference file** dialog which appears, followed by a *click* on the dialog's **Open** button. This brings up the **Attach External Reference** dialog (Fig. 10.16) showing **Fig13** in its **Name** field. *Click* the dialog's **OK** button.

5. The spindle drawing appears on screen ready to be *dragged* into position. Place it in position as indicated in Fig. 10.17.

Fig. 10.12 Example – **External references** – original drawing.

Fig. 10.13 The spindle drawing saved as **Fig13.dwg**.

Fig. 10.14 The **External Reference** tool in the **View/Palettes** panel.

Fig. 10.15 The **External References** palette.

Fig. 10.16 The **Attach External Reference** dialog.

**Fig. 10.17** The spindle in place in the original drawing.

6. Save the drawing with its Xref to its original file name.
7. Open **Fig15.dwg** and make changes as shown in Fig. 10.18.
8. Now reopen the original drawing. The **external reference** within the drawing has changed in accordance with the alterations to the spindle drawing. Fig. 10.19 shows the changes in the front view of the original drawing.

**Fig. 10.18** The revised **spindle.dwg** drawing.　　**Fig. 10.19** Example – **Xrefs.**

NOTE →

In this example to ensure accuracy of drawing, the **external reference** will need to be exploded and parts of the spindle changed to hidden detail lines.

DGNIMPORT AND DGNEXPORT

Drawings constructed in MicroStation V8 format (***.dgn**) can be imported into AutoCAD 2012 format using the command **dgnimport** at the command line. AutoCAD drawings in AutoCAD 2004 format can be exported into MicroStation ***.dgn** format using the command **dgnexport**.

## EXAMPLE OF IMPORTING A *.DGN DRAWING INTO AUTOCAD

1. Fig. 10.20 shows an example of an orthographic drawing constructed in MicroStation V8.
2. In AutoCAD 2012 at the command line *enter* **dgnimport**. The dialog (Fig. 10.21) appears on screen from which the required drawing file name can be selected.

**Fig. 10.20** Example – a drawing in MicroStation V8.

**Fig. 10.21** The **Import DGN File** dialog.

When the **Open** button of the dialog is *clicked* a warning window appears informing the operator of steps to take in order to load the drawing. When completed the drawing loads (Fig. 10.22).

In a similar manner AutoCAD drawing files can be exported to MicroStation using the command **dgnexport** *entered* at the command line.

**Fig. 10.22** The **\*.dgn** file imported into AutoCAD 2012.

## MULTIPLE DESIGN ENVIRONMENT

1. Open several drawings in AutoCAD; in this example four separate drawings have been opened.
2. In the **View/Windows** panel *click* **Tile Horizontally** (Fig. 10.23).

**Fig. 10.23** Selecting **Tile Horizontally** from the **View/Windows** panel.

The four drawings rearrange as shown in Fig. 10.24.

**Fig. 10.24** Four drawings in the **Multiple Design Environment**.

## REVISION NOTES

1. The Edit tools Copy Clip, Copy with Base Point, and Copy Link enable objects from AutoCAD 2012 to be copied for Pasting into other applications.

2. Objects can be copied from other applications to be pasted into the AutoCAD 2012 window.

3. Drawings saved in AutoCAD as DXF (*.dxf) files can be opened in other Computer Aided Design (CAD) applications.

4. Similarly drawings saved in other CAD applications as *.dxf files can be opened in AutoCAD 2012.

5. Raster files of the format types *.bmp, *.jpg, *pcx, *.tga, *.tif, among other raster type file objects, can be inserted into AutoCAD 2012 drawings.

6. Drawings saved to the Encapsulated Postscript (*.eps) file format can be inserted into documents of other applications.

7. Changes made in a drawing saved as an *.eps file will be reflected in the drawing inserted as an *.eps file in another application.

8. When a drawing is inserted into another drawing as an external reference, changes made to the inserted drawing will be updated in the drawing into which it has been inserted.

9. A number of drawings can be opened at the same time in the AutoCAD 2012 window.

10. Drawings constructed in MicroStation V8 can be imported into AutoCAD 2012 using the command dgnimport.

11. Drawings constructed in AutoCAD 2012 can be saved as MicroStation *.dgn drawings to be opened in MicroStation V8.

**10** **PART A** 2D Design **EXERCISES**

## EXERCISES

Methods of constructing answers to the following exercises can be found in the free website:

**http://www.routledge.com/cw/yarwood**

1.  Fig. 10.25 shows a pattern formed by inserting an **external reference** and then copying or arraying the **external reference**.

    The hatched parts of the **external reference** drawing were then changed using a different hatch pattern. The result of the change in the hatching is shown in Fig. 10.26.

    Construct a similar **Xref** drawing, insert as an **Xref**, array or copy to form the pattern, then change the hatching, save the **Xref** drawing and note the results.

**Fig. 10.25** Exercise 1 – original pattern.

**Fig. 10.26** Exercise 1 – Fig.10.25 after change in hatching.

2.  Fig. 10.27 shows a rendering of a roller between two end holders. Fig. 10.28 gives details of the end holders and the roller in orthographic projections.

**Fig. 10.27** Exercise 2 – a rendering of the holders and roller.

**Fig. 10.28** Exercise 2 – details of the parts of the holders and roller.

Construct a full size front view of the roller and save to a file name **roller.dwg**. Then as a separate drawing construct a front view of the two end holders in their correct positions to receive the roller and save to the file name **assembly.dwg**.

Open the **roller.dwg** and change its outline as shown in Fig. 10.29. Save the drawing. Open the **assembly.dwg** and note the change in the inserted **Xref**.

**Fig. 10.29** Exercise 2 – The amended **Xref** drawing.

3. *Click* **Image...** in the **Reference** panel and insert a **JPEG** image (*.jpg* file) of a photograph into the AutoCAD 2010 window. An example is given in Fig. 10.30.

**Fig. 10.30** Exercise 3 – example.

4. Using **Copy** from the **Insert** drop-down menu, copy a drawing from AutoCAD 2010 into a Microsoft Word document. An example is given in Fig. 10.31. Add some appropriate text.

10 **PART A** 2D Design EXERCISES

**5.** The plan in Figs 10.1 and 10.2 is incorrect in that some details have been missed from the drawing. Can you identify the error?

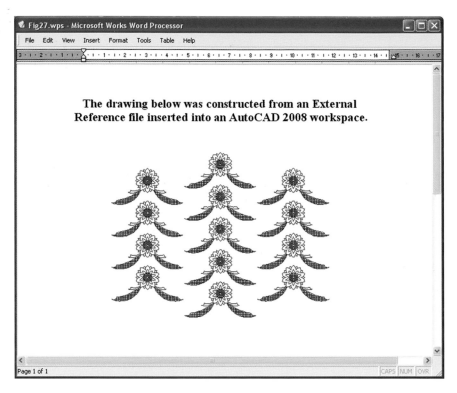

**Fig. 10.31** Exercise 4 – an example.

# SHEET SETS

## AIMS OF THIS CHAPTER

The aims of this chapter are:

1. To introduce sheet sets.
2. To describe the use of the **Sheet Set Manager**.
3. To give an example of a sheet set based on a design of a two-storey house.

## SHEET SETS

When anything is to be manufactured or constructed, whether it be a building, an engineering design, an electronics device, or any other form of manufactured artifact, a variety of documents, many in the form of technical drawings, will be needed to convey to those responsible for constructing the design all the information necessary to be able to proceed according to the wishes of the designer. Such sets of drawings may be passed between the people or companies responsible for the construction, enabling all those involved to make adjustments or suggest changes to the design. In some cases there may well be a considerable number of drawings required in such sets of drawings. In AutoCAD 2012 all the drawings from which a design is to be manufactured can be gathered together in a **sheet set**. This chapter shows how a much reduced sheet set of drawings for the construction of a house at 64 Pheasant Drive can be produced. Some other drawings, particularly detail drawings, would be required in this example, but to save page space, the sheet set described here consists of only four drawings with a subset of another four drawings.

---

Introduction to AutoCAD 2012. ISBN: 978-0-08-096947-3

## SHEET SET FOR 64 PHEASANT DRIVE

1. Construct a template **64 Pheasant Drive.dwt** based upon the acadiso.dwt template, but including a border and a title block. Save the template in a **Layout1** format. An example of the title block from one of the drawings constructed in this template is shown in Fig. 11.1.

Scale:
1:50

Date:
12:09:07

Drawing No:
2

Title:
Building plan

**Fig. 11.1** The title block from Drawing number **2** of the sheet set drawings.

2. Construct each of the drawings which will form the sheet set in this drawing template. The whole set of drawings is shown in Fig. 11.2. Save the drawings in a directory; in this example the name that has been given is **64 Pheasant Drive**.

**Fig. 11.2** The eight drawings in the **64 Pheasant Drive** sheet set.

3. *Click* **Sheet Set Manager** in the **View/Palettes** panel (Fig. 11.3).

**Fig. 11.3** Selecting **Sheet Set Manager** from the **View/Palettes** panel.

The **Sheet Set Manager** palette appears (Fig. 11.4).

**Fig. 11.4** The **Sheet Set Manager** palette.

*Click* **New Sheet Set…** in the popup menu at the top of the palettes. The first of a series of **Create Sheet Set** dialogs appears – the **Create Sheet Set – Begin** dialog (Fig. 11.5).

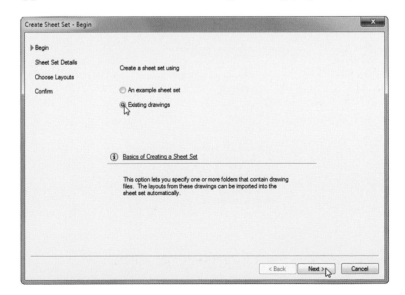

**Fig. 11.5** The first of the **Create Sheet Set** dialogs: **Create Sheet Set – Begin**.

*Click* the radio button next to **Existing drawings,** followed by a *click* on the **Next** button – the next dialog **Sheet Set Details** appears (Fig. 11.6).

**Fig. 11.6** The **Sheet Set Details** dialog.

4. *Enter* details as shown in the dialog in Fig. 11.6. Then *click* the **Next** button to bring the **Choose Layouts** dialog to screen (Fig. 11.7). *Click* its **Browse** button and from the **Browse for Folder** list which comes to screen, *pick* the directory **64 Pheasant Drive.** *Click* the **OK** button and the drawings held in the directory appear in the **Choose Layouts** dialog (Fig. 11.7).

**Fig. 11.7** The **Choose Layouts** dialog.

If satisfied the list is correct, *click* the **Next** button. A **Confirm** dialog appears (Fig. 11.8).

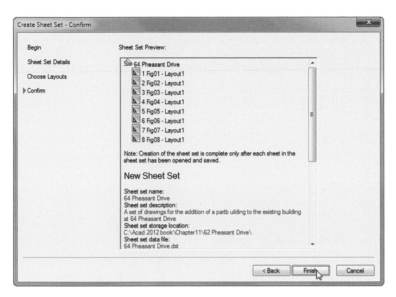

**Fig. 11.8** The Confirm dialog.

If satisfied *click* the **Finish** button and the **Sheet Set Manager** palette appears showing the drawings which will be in the **64 Pheasant Drive** sheet set (Fig. 11.9).

**Fig. 11.9** The **Sheet Set Manager** palette for **64 Pheasant Drive.**

## NOTES →

1. The eight drawings in the sheet set are shown in Fig. 11.9. If any of the drawings in the sheet set are subsequently amended or changed, the drawings will include any changes or amendments when they are opened again from the **64 Pheasant Drive Sheet Manager** palette.

→ NOTES (CONTINUED)

2. Drawings can only be placed into sheet sets if they have been saved in a **Layout** screen. Note that all of the drawings shown in the **64 Pheasant Drive Sheet Set Manager** have **Layout1** after the drawing names because each has been saved after being constructed in a **Layout1** template.

3. Sheet sets in the form of **DWF** (Design Web Format) files can be sent via email to others who are using the drawings, or placed on an intranet. The method of producing a **DWF** for the **64 Pheasant Drive** Sheet Set follows.

## 64 PHEASANT DRIVE DWF

1. In the **64 Pheasant Drive Sheet Set Manager** *click* the **Publish** icon, followed by a *click* on **Publish to DWF** in the menu; which will appear as shown in Fig. 11.10.

**Fig. 11.10** The **Publish** icon in the **Sheet Set Manager.**

The **Specify DWF File** dialog appears (Fig. 11.11). *Enter* **64 Pheasant Drive** in the **File name** field followed by a *click* on the **Select** button. A warning window (Fig. 11.12) appears. *Click* its **Close** button. In Fig. 11.10 the **Publish** icon is in the **Sheet Set Manager** button. The **Publish Job in Progress** icon in the bottom right-hand corner of the AutoCAD 2012 window starts fluctuating in shape showing that the DWF file is being processed (Fig. 11.12). When the icon becomes stationary, *right-click* on it, and then *click* on **View Plotted File**. This *right-click* menu, including **View Plotted File,** is shown in Fig. 11.13.

**Fig. 11.11** The **Specify DWF File** dialog.

**Fig. 11.12** The **Publish Job in Progress** icon.

**Fig. 11.13** The *right-click* menu of the icon.

2. The **Autodesk Design Review** window appears showing the
   **64 Pheasant Drive.dwf** file (Fig. 11.14). *Click* on the arrow
   **Next Page (Page on)** to see other drawings in the DWF file.
3. If required the Design Review file can be sent between people
   by email as an attachment, opened in a company's intranet or,
   indeed, included within an internet web page.

Fig. 11.14 The **Autodesk Design Review** showing details of the **64 Pheasant Drive.dwf** file.

## REVISION NOTES

1. To start off a new sheet set, select the Sheet Set Manager icon in the Tools/Palettes panel
2. Sheet sets can only contain drawings saved in Layout format.
3. Sheet sets can be published as Design Review Format (*.dwf) files which can be sent between offices by email, published on an intranet, or published on a web page.
4. Subsets can be included in sheet sets.
5. Changes or amendments made to any drawings in a sheet set are reflected in the sheet set drawings when the sheet set is opened.

# EXERCISES

Methods of constructing answers to the following exercises can be found in the free website:

**http://www.routledge.com/cw/yarwood**

**1.** Fig. 11.15 shows an exploded orthographic projection of the parts of a piston and its connecting rod. There are four parts in the assembly. Small drawings of the required sheet set are shown in Fig. 11.17.

Construct the drawing in Fig. 11.15 and also the four drawings of its parts. Save each of the drawings in a **Layout1** format and construct the sheet set which contains the five drawings.

**Fig. 11.15** Exercise 1 – exploded orthographic projection.

**Fig.11.16** The **DWF** for Exercise 1.

**Fig. 11.17** Exercise 1 – the five drawings in the sheet set.

Construct the **DWF** file of the sheet set. Experiment by sending it to a friend via email as an attachment to a document, asking him/her to return the whole email to you without changes. When the email is returned, open its DWF file and *click* each drawing icon in turn to check the contents of the drawings.

**2.** Construct a similar sheet set as in the answer to Exercise 1 from the exploded orthographic drawing of a **Machine adjusting spindle** given in Fig. 11.18.

**Fig. 11.18** Exercise 2.

# PART B

## 3D DESIGN

CHAPTER **12**

# INTRODUCING 3D MODELING

## AIMS OF THIS CHAPTER

The aims of this chapter are:

1. To introduce the tools used for the construction of 3D solid models.
2. To give examples of the construction of 3D solid models using tools from the **Home/Create** panel.
3. To give examples of 2D outlines suitable as a basis for the construction of 3D solid models.
4. To give examples of constructions involving the Boolean operators – **Union, Subtract,** and **Intersect.**

## INTRODUCTION

As shown in Chapter 1 the AutoCAD coordinate system includes a third coordinate known as direction **Z**. This third coordinate is not utilized when dealing with the 2D drawing outlined in previous chapters. 3D model drawings make use of this third **Z** coordinate.

## THE 3D BASICS WORKSPACE

It is possible to construct 3D model drawings in the **2D Drafting & Annotation** workspaces; but in **Part B** of this book we will be working in either the **3D Basics** or in the **3D Modeling** workspaces. To set the first of these workspaces *click* the **Workspace Settings** icon in the status bar and select **3D Introduction** from the menu which appears (Fig. 12.1).

**Fig. 12.1** Selecting **3D Basics** from the **Workspace Switching** menu.

The workspace in Fig. 12.2 is the window in which the examples in this chapter will be constructed.

**Fig. 12.2** The **3D Basics** workspace.

## METHODS OF CALLING TOOLS FOR 3D MODELING

The default panels of the **3D Basics** ribbon are shown in Fig. 12.3.

**Fig. 12.3** The default **3D Basics** panels.

When calling the tools for the construction of 3D model drawings 3D tools can be called by:

1. A *click* on a tool icon in a **3D Basics** panel.
2. *Entering* the tool name at the command line followed by pressing the *Return* button of the mouse or the **Return** key of the keyboard.
3. Some of the 3D tools have an abbreviation which can be *entered* at the command line instead of its full name.
4. Using the **Dynamic Input** method.

## NOTES →

1. Most operators will use a variation of four different methods; no matter which method is being used, the result of calling a tool results in prompt sequences appearing at the command prompt (or if using **Dynamic Input** on screen) as in the following example:

```
Command: enter box right-click
Specify first corner or [Center]: enter 90,120
  right-click
Specify other corner or [Cube/Length]: enter
  150,200
Specify height or [2Point]: enter 50
Command:
```

Or, if the tool is called from its tool icon, or from a drop-down menu:

```
Command: _box
Specify first corner or [Center]: enter 90,120
  right-click
Specify other corner or [Cube/Length]: enter
  150,200
Specify height or [2Point]: enter 50
Command:
```

This command is the same as that used in 2D drawing.

2. In the following pages, if the tool's sequences are to be repeated, they may be replaced by an abbreviated form such as:

```
Command: box
[prompts]: 90,120
[prompts]: 150,200
```

→ NOTES (CONTINUED)

3. The examples shown in this chapter will be based on layers set as follows:

a) *Click* the **Layer Properties** icon in the **Home/Layers** panel (Fig. 12.4).

**Fig. 12.4** The **Layer Properties** icon in the **Home/Layers** panel.

b) In the **Layer Properties Manager** which appears make settings as shown in Fig. 12.5.

**Fig. 12.5** The settings in the **Layer Properties Manager**.

## THE POLYSOLID TOOL (FIG. 12.8)

1. Set layer **Blue** as the current layer.
2. Construct an octagon of edge length **60** using the **Polygon** tool.
3. *Click* **SW Isometric** in the **Layers & View** panel (Fig. 12.6).

**Fig. 12.6** Selecting **SW Isometric** from the **3D navigation** drop-down menu in the **Layers & View** panel.

**4.** Call the **Polysolid** tool from the **Home/Create panel** (Fig. 12.7).

The command line shows:

```
Command: _Polysolid Height = 0, Width = 0,
 Justification = Center
Specify start point or [Object/Height/Width/Justify]
 <Object>: enter h right-click
Specify height <0>: enter 60 right-click
Height = 60, Width = 0, Justification = Center
Specify start point or [Object/Height/Width/Justify]
 <Object>: enter w right-click
Specify width <0>: 5
Height = 60, Width = 5, Justification = Center
Specify start point or [Object/Height/Width/Justify]
 <Object>: pick the polygon
Select object: right-click
Command:
```

And the **Polysolid** forms.

**5.** Select **Conceptual** from the **Layers & View** panel (Fig. 12.8)

The result is shown in Fig. 12.9.

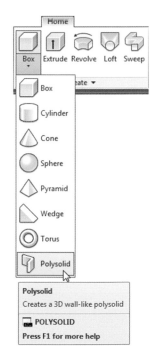

**Fig. 12.7** The **Polysolid** tool icon in the **Home/Create** panel.

Fig. 12.9 The **Polysolid** tool example.

Fig. 12.8 Selecting **Conceptual** shading from **Visual Styles** in the **Layers & View** panel.

## 2D OUTLINES SUITABLE FOR 3D MODELS

When constructing 2D outlines suitable as a basis for constructing some forms of 3D model, select a tool from the **Home/Draw** panel, or *enter* tool names or abbreviations for the tools at the command line. If constructed using tools such as **Line, Circle,** and **Ellipse,** before being of any use for 3D modeling, outlines must be changed into regions with the **Region** tool. Closed polylines can be used without the need to use the **Region** tool.

## EXAMPLE – OUTLINES & REGION (FIG. 12.10)

1. Construct the top left-hand drawing of Fig. 12.10 using the **Line and Circle** tools.
2. *Enter* **region** or **reg** at the command line. The command line shows:

```
Command: _region
Select objects: window the left-hand rectangle 1
 found
Select objects: right-click
1 loop extracted.
1 Region created.
Command:
```

1.3 regions  3. Union of 2 unions

2. Union of 3 regions  4. Subtract region from Union

**Fig. 12.10** Example – **Line and Circle** outlines and **Region**.

And the **Line** outline is changed to a region. Repeat for the circle and the right-hand rectangle. Three regions will be formed.

3. Drawing **2** – call the **Union** tool from the **Home/Edit** panel (Fig. 12.11).

**Fig. 12.11** Selecting the **Union** tool from the **Home/Edit** panel.

The command line shows:

```
Command: _union
Select objects: pick the left-hand region 1 found
Select objects: pick the circular region 1 found,
 2 total
Select objects: pick the right-hand region 1 found,
 3 total
Select objects: right-click
Command:
```

4. Drawing **3** – with the **Union** tool form a union of the left-hand region and the circular region.

5. Drawing **4** – call the **Subtract** tool, also from the **Home/Edit** panel. The command line shows:

```
Command: _subtract Select solids and regions to
  subtract from...
Select objects: pick the region just formed 1 found
Select objects: right-click
Select solids and regions to subtract...
Select objects: pick the right-hand region 1 found
Select objects: right-click
Command:
```

## THE EXTRUDE TOOL

The **Extrude** tool can be called with a *click* on its name in the **Home/Create** panel (Fig. 12.12), or by *entering* **extrude** or its abbreviation **ext** at the command line.

### EXAMPLES OF THE USE OF THE EXTRUDE TOOL

The first two examples of forming regions given in Figs 12.10 and 12.11 are used to show results of using the **Extrude** tool.

### FIRST EXAMPLE – EXTRUDE (FIG. 12.13)

From the first example of forming a region:

1. Open Fig. 12.10. Erase all but the region **2**.
2. Make layer **Green** current.
3. Call **Extrude** (Fig. 12.12).

**Fig. 12.12** The **Extrude** tool from the **Home/Create** panel.

The command line shows:

```
Command: _extrude
Current wire frame density: ISOLINES=4
Closed profiles creation mode = Solid
```
**Select objects to extrude or [MOde]:** *pick* region 1
  found

```
Select objects to extrude or [MOde]: right-click
Specify height of extrusion or [Direction/Path/Taper
 angle/Expression] <45>: enter 50 right-click
Command:
```

4. Place in the **Layers & View/3D Navigation/SW/Isometric** view.
5. Call **Zoom** and zoom to **1**.
6. Place in **Visual Style/Conceptual**

The result is shown in Fig. 12.13.

**Fig. 12.13** First example – **Extrude.**

## NOTES →

1. In the above example we made use of an isometric view possible from the **3D Navigation** drop-down menu in the **Home/Layers & Views** panel (Fig.12.6). The **3D Navigation** drop-down menu allows a model to be shown from a variety of different angles.

2. Note the **Current wire frame density: ISOLINES = 4** in the prompts sequence when **Extrude** is called. The setting of **4** is suitable when extruding plines or regions consisting of straight lines; but when arcs are being extruded it may be better to set **ISOLINES** to a higher figure as follows:

```
Command: enter isolines right-click
Enter new value for ISOLINES <4>: enter 16
 right-click
Command:
```

3. Note the prompt [**MOde**] in the line

```
Select objects to extrude or [MOde]:
```

If **mo** is *entered* as a response to this prompt line, the following prompts appear:

```
Closed profiles creation mode[SOlid/SUrface]
 <Solid>: _SO
```

This allows the extrusion to be shown in solid or surface format.

# SECOND EXAMPLE – EXTRUDE (FIG. 12.14)

1. Open Fig. 12.10 and erase all but the region **3**.
2. Make the layer **Blue** current.
3. Set **ISOLINES** to **16**.
4. Call the **Extrude** tool. The command line shows:

```
Command: _extrude
Current wire frame density: ISOLINES=4, Closed
 profiles creation mode = Solid
Select objects to extrude or [MOde]: _MO Closed
 profiles creation mode
[SOlid/SUrface] <Solid>: _SO
Select objects to extrude or [MOde]: pick the region
 3 1 found
Select objects to extrude or [MOde]:
Specify height of extrusion or [Direction/Path/Taper
 angle/Expression]: enter t right-click
Specify angle of taper for extrusion or [Expression]
 <0>: enter 10 right-click
Specify height of extrusion or [Direction/Path/Taper
 angle/Expression]: enter 100 right-click
Command:
```

**Fig. 12.14** Second example – **Extrude**.

5. In the **Layers & View/3D Navigation** menu select **NE Isometric**.
6. **Zoom** to **1**.
7. Place in **Visual Styles/Hidden**.

# THIRD EXAMPLE – EXTRUDE (FIG. 12.16)

1. Make layer **Magnolia** current.
2. Construct an **80 x 50** rectangle, filleted to a radius of **15**. Then in the **3D Navigation/Front** view and using the **3D Polyline** tool from the **Home/Draw** panel (Fig. 12.15) construct 3 3D polylines each of length **45** and at **45°** to each other at the center of the outline as shown in Fig. 12.16.
3. Place the screen in the **3D Navigation/SW Isometric** view.
4. Set **ISOLINES** to **24**.
5. Call the **Extrude** tool. The command line shows:

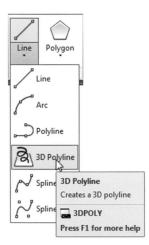

**Fig. 12.15** The **3D Polyline** tool from the **Home/Draw** panel.

```
Command: _extrude
Current wire frame density: ISOLINES=24, Closed
 profiles creation mode = Solid
Select objects to extrude or [MOde]: _MO Closed
 profiles creation mode
[SOlid/SUrface] <Solid>: _SO
```

```
Select objects to extrude or [MOde]: pick the
  rectangle 1 found
Select objects to extrude or [MOde]: right-click
Specify height of extrusion or [Direction/Path/Taper
  angle/Expression]: enter t right-click
```
**Select extrusion path or [Taper angle]:** *pick* path
  *right-click*
```
Command:
```

**6.** Place the model in **Visual Styles/Realistic.**

The result is shown in Fig. 12.16.

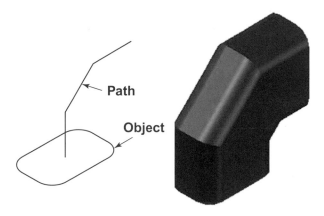

**Fig. 12.16** Second example – **Extrude.**

## THE REVOLVE TOOL

The **Revolve** tool can be called with a *click* on its tool icon in the **Home/Create** panel, by a *click* or by *entering* **revolve** or its abbreviation **rev** at the command line.

## EXAMPLES – REVOLVE TOOL

Solids of revolution can be constructed from closed plines or from regions.

### FIRST EXAMPLE – REVOLVE (12.19)

**1.** Construct the closed polyline in Fig. 12.17.
**2.** Make layer **Red** current.
**3.** Set **ISOLINES** to 24.
**4.** Call the **Revolve** tool from the **Home/Create** panel (Fig. 12.18).

**Fig. 12.17** First example – **Revolve**. The closed pline.

**Fig. 12.18** The **Revolve** tool from the **Home/Create** panel.

The command line shows:

```
Command: _revolve
Current wire frame density: ISOLINES=4, Closed
 profiles creation mode = Solid
Select objects to revolve or [MOde]: _MO Closed
 profiles creation mode[SOlid/SUrface] <Solid>: _SO
Select objects to revolve or [MOde]: Pick the pline
 1 found
Select objects to revolve or [MOde]: right-click
Specify axis start point or define axis by [Object/
 X/Y/Z] <Object>: pick
Specify axis endpoint: pick
Specify angle of revolution or [STart angle/Reverse/
 Expression] <360>: right-click
Command:
```

5. Place in the **3D Navigation/NE Isometric** view. **Zoom** to **1**.
6. Shade with **Visual Styles/Shaded**.

The result is shown in Fig. 12.19.

Fig. 12.19 First example – **Revolve**.

## SECOND EXAMPLE – REVOLVE (FIG. 12.21)

1. Make layer **Yellow** current.
2. Place the screen in the **3D Navigate/Front** view. **Zoom** to **1**
3. Construct the pline outline (Fig. 12.20).
4. Set **ISOLINES** to **24**.
5. Call the **Revolve** tool and construct a solid of revolution.
6. Place the screen in **3D Navigate/SW Isometric**. **Zoom** to **1**.
7. Place in **Visual Styles/Shades of Gray**.

Semi-ellipse based
on 180 x 100 axes

Fig. 12.20 Second example
– **Revolve**. The pline outline.

Fig. 12.21 Second example – **Revolve**.

## THIRD EXAMPLE – REVOLVE (FIG. 12.22)

1. Make **Green** the current layer.
2. Place the screen in the **3D Navigate/Front** view.
3. Construct the pline (left-hand drawing of Fig. 12.22). The
   drawing must be either a closed pline or a region.

4. Set **Isolines** to **24**.
5. Call **Revolve** and form a solid of revolution through **180°**.
6. Place the model in the **3D Navigate/NE Isometric. Zoom** to **1**.
7. Place in **Visual Styles/Conceptual**.

The result is shown in Fig. 12.22 (right-hand drawing).

**Fig. 12.22** Third example – **Revolve**. The outline to be revolved and the solid of revolution.

## OTHER TOOLS FROM HOME/CREATE

## FIRST EXAMPLE – BOX (FIG. 12.24)

1. Make **Magenta** the current layer.
2. Place the window in the **3D Navigate/Front** view.
3. Set **Isolines** to **4**.
4. *Click* the **Box** tool icon in the **Home/Create** panel (Fig. 12.23).

The command line shows:

```
Command: _box
Specify first corner or [Center]: enter 90,90
 right-click
Specify other corner or [Cube/Length]: enter
 110,-30 right-click
Specify height or [2Point]: enter 75 right-click
Command: right-click
BOX Specify first corner or [Center]: 110,90
Specify other corner or [Cube/Length]: 170,70
Specify height or [2Point]: 75
```

**Fig. 12.23** Selecting **Box** from the **Home/Create** panel.

```
Command:
BOX Specify first corner or [Center]: 110,-10
Specify other corner or [Cube/Length]: 200,-30
Specify height or [2Point]: 75
Command:
```

5. Place in the **ViewCube/Isometric** view. **Zoom to 1**.
6. Call the **Union** tool from the **Home/Edit** panel. The command line shows:

```
Command: _union
Select objects: pick one of the boxes 1 found
Select objects: pick the second of box 1 found, 2
 total
Select objects: pick the third box 1 found, 3 total
Select objects: right-click
Command:
```

**Fig. 12.24** First example – Box.

And the three boxes are joined in a single union.

7. Place in **Visual Styles/Conceptual**.

## SECOND EXAMPLE – SPHERE AND CYLINDER (FIG. 12.25)

1. Make layer **Green** current.
2. Set **ISOLINES** to **16**.
3. *Click* the **Sphere** tool icon from the **Home/Create** panel. The command line shows:

```
Command: _sphere
Specify center point or [3P/2P/Ttr]: 180,170
Specify radius or [Diameter]: 50
Command:
```

4. *Click* the **Cylinder** tool icon in the **Home/Create** panel. The command line shows:

```
Command: _cylinder
Specify center point of base or [3P/2P/Ttr/
 Elliptical]: 180,170
Specify base radius or [Diameter]: 25
Specify height or [2Point/Axis endpoint]: 110
Command:
```

5. Place the screen in the **3D Navigate/Front** view. **Zoom to 1**.
6. With the **Move** tool (from the **Home/Modify** panel), move the cylinder vertically down so that the bottom of the cylinder is at the bottom of the sphere.

**7.** *Click* the **Subtract** tool icon in the **Home/Edit** panel. The command line shows:

```
Command: _subtract Select solids and regions to
  subtract from...
Select objects: pick the sphere 1 found
Select objects: right-click
Select solids and regions to subtract
Select objects: pick the cylinder 1 found
Select objects: right-click
Command:
```

**8.** Place the screen in **3D Navigate/SW Isometric**. **Zoom** to **1**.
**9.** Place in **Visual Styles/Conceptual**.

The result is shown in Fig. 12.25.

**Fig. 12.25** Second example – **Sphere and Cylinder**.

## THIRD EXAMPLE – CYLINDER, CONE, AND SPHERE (FIG. 12.26)

**1.** Make **Blue** the current layer.
**2.** Set **Isolines** to **24**.
**3.** Place in the **3D Navigate/Front** view.
**4.** Call the **Cylinder** tool and with a center 170,150 construct a cylinder of radius **60** and height **15**.
**5.** *Click* the **Cone** tool in the **Home/Create** panel. The command line shows:

```
Command: _cone
Specify center point of base or [3P/2P/Ttr/
  Elliptical]: 170,150
Specify base radius or [Diameter]: 40
Specify height or [2Point/Axis endpoint/Top radius]:
  150
Command:
```

6. Call the **Sphere** tool and construct a sphere of center **170,150** and radius **45**.

7. Place the screen in the **3D Navigate/Front** view and with the **Move** tool, move the cone and sphere so that the cone is resting on the cylinder and the center of the sphere is at the apex of the cone.

8. Place in the **3D Navigate/SW Isometric** view, **Zoom** to **1** and with **Union** form a single 3D model from the three objects.

9. Place in **Visual Styles/Shades of Gray**.

The result is shown in Fig. 12.26.

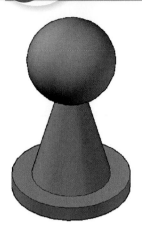

**Fig. 12.26** Third example – **Cylinder**, **Cone**, and **Sphere**.

## FOURTH EXAMPLE – BOX, AND WEDGE (FIG. 12.27)

1. Make layer **Cyan** current.

2. Place in the **3D Navigate/Top** view.

3. *Click* the **Box** tool icon in the **Home/Create** panel and construct two boxes, the first from corners **70,210** and **290,120** of height **10**, the second of corners **120,200,10** and **240,120,10** and of height **80**.

4. Place the screen in the **3D Navigate/Front** view and **Zoom** to **1**.

5. *Click* the **Wedge** tool icon in the **Home/Create** panel. The command line shows:

```
Command: _wedge
Specify first corner or [Center]: 120,170,10
Specify other corner or [Cube/Length]: 80,160,10
Specify height or [2Point]: 70
Command: right-click
WEDGE
Specify first corner of wedge or [Center]:
 240,170,10
Specify corner or [Cube/Length]: 280,160,10
Specify height or [2Point]: 70
Command:
```

6. Place the screen in **3D Navigate/SW Isometric** and **Zoom** to **1**.

7. Call the **Union** tool from the **Home/Edit** panel; and in response to the prompts in the tool's sequences *pick* each of the 4 objects in turn to form a union of the 4 objects.

8. Place in **Visual Styles/Conceptual**

The result is shown in Fig. 12.27.

**Fig. 12.27** Fourth example – **Box** and **Wedge**.

## FIFTH EXAMPLE – CYLINDER AND TORUS (FIG. 12.28)

1. Make layer **Red** current.
2. Set **Isolines** to **24**.
3. Using the **Cylinder** tool from the **Home/Create** panel, construct a cylinder of center **180,160**, of radius **40** and height **120**.
4. *Click* the **Torus** tool icon in the **Home/Create** panel. The command line shows:

```
Command: _torus
Specify center point or [3P/2P/Ttr]: 180,160,10
Specify radius or [Diameter]: 40
Specify tube radius or [2Point/Diameter]: 10
Command: right-click
TORUS
Specify center point or [3P/2P/Ttr]: 180,160,110
Specify radius or [Diameter] <40>: right-click
Specify tube radius or [2Point/Diameter] <10>:
 right-click
Command:
```

5. Call the **Cylinder** tool again and construct another cylinder of center **180,160**, of radius **35** and height **120**.
6. Place in the **3D Navigate/SW Isometric** view and **Zoom** to **1**.
7. *Click* the **Union** tool icon in the **Home/Edit** panel and form a union of the larger cylinder and the two tori.
8. *Click* the **Subtract** tool icon in the **Home/Edit** panel and subtract the smaller cylinder from the union.
9. Place in **Visual Styles/X-Ray**.

The result is shown in Fig. 12.28.

**Fig. 12.28** Fifth example – **Cylinder** and **Torus**.

## THE FILLET AND CHAMFER TOOLS

## EXAMPLE – FILLET AND CHAMFER (FIG. 12.33)

1. Set layer **Green** as the current layer.
2. Set **Isolines** to **16**.
3. Working to the sizes given in Fig. 12.29 and using the **Box** and **Cylinder** tools, construct the 3D model in Fig. 12.30.
4. Place in the **3D Navigate/SW Isometric** view. **Union** the two boxes and with the **Subtract tool**; subtract the cylinders from the union.

Box 160 × 100 × 10     Elliptical cylinder
                       80 × 40 height 60

Cylinders R5    Box 120 × 60 × 50
height 10

**Fig. 12.29** Example – **Fillet and Chamfer** – sizes for the model.

**Fig. 12.30** Example – **Fillet and Chamfer** – isometric view – the model before using the tools.

---

NOTE ➜

To construct the elliptical cylinder, call the **Cylinder** tool from the **Home/Modeling** panel. The command line shows:

```
Command: _cylinder
Specify center point of base or [3P/2P/Ttr/
  Elliptical]: enter e
  right-click
Specify endpoint of first axis or [Center]: 130,160
Specify other endpoint of first axis: 210,160
Specify endpoint of second axis: 170,180
Specify height or [2Point/Axis endpoint]: 50
Command:
```

5. *Click* the **Fillet** tool icon in the **Home/Modify** panel (Fig. 12.31).

**Fig. 12.31** The **Fillet** tool icon in the **Home/Modify** panel flyout.

The command line shows:

```
Command: _fillet
Current settings: Mode = TRIM. Radius = 0
Specify first object or [Undo/Polyline/Radius/Trim/
 Multiple]: enter r (Radius) right-click
Specify fillet radius <0>: 10
Select first object: pick one corner
Select an edge or [Chain/Radius]: pick a second
 corner
Select an edge or [Chain/Radius]: pick a third
 corner
Select an edge or [Chain/Radius]: pick the fourth
 corner
Select an edge or [Chain/Radius]: right-click
4 edge(s) selected for fillet.
Command:
```

**6.** *Click* the **Chamfer** tool in the **Home/Modify** panel (Fig. 12.32).

**Fig. 12.32** The **Chamfer** tool icon in the **Home/Modify** panel flyout.

The command line shows:

```
Command: _chamfer
(TRIM mode) Current chamfer Dist1 = 0, Dist2 = 0
```

```
Select first line or [Undo/Polyline/Distance/Angle/
 Trim/mEthod/Multiple]: enter d right-click
Specify first chamfer distance <0>: 10
Specify second chamfer distance <10>:
Select first line or [Undo/Polyline/Distance/Angle/
 Trim/mEthod/Multiple]: pick one corner One side of
 the box highlights
Base surface selection...
Enter surface selection option [Next/OK (current)]
 <OK>: right-click
Specify base surface chamfer distance <10>:
 right-click
Specify other surface chamfer distance <10>:
 right-click
Select an edge or [Loop]: pick the edge
Select an edge or [Loop]: pick the second edge
Select an edge or [Loop]: right click
Command:
```

And the edges is chamfered. Repeat to chamfer the other three edges.

**7.** Place in **Visual Styles/Shaded with Edges**.

Fig. 12.33 shows the completed 3D model.

**Fig. 12.33** Example – **Fillet** and **Chamfer**.

## NOTE ON THE TOOLS UNION, SUBTRACT, AND INTERSECT →

The tools **Union**, **Subtract**, and **Intersect** found in the **Home/Edit** panel are known as the **Boolean** operators after the mathematician George Boole. They can be used to form unions, subtractions, or intersection between extrusions, solids of revolution, or any of the 3D Objects.

## CONSTRUCTING 3D SURFACES USING THE EXTRUDE TOOL

In this example of the construction of a 3D surface model the use of the **Dynamic Input** (DYN) method of construction will be shown.

1. Place the AutoCAD drawing area in the **3D Navigation/SW Isometric** view.
2. *Click* the **Dynamic Input** button in the status bar to make dynamic input active.

## EXAMPLE – DYNAMIC INPUT (FIG. 12.35)

1. Using the **Line** tool from the **Home/Draw** panel construct the outline (Fig. 12.34).

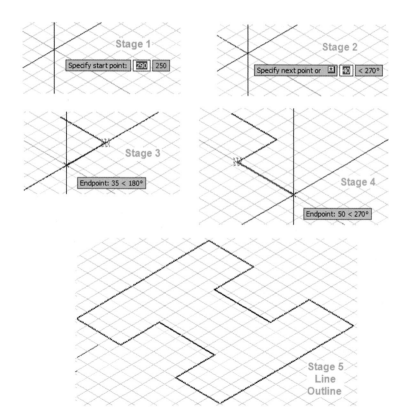

**Fig. 12.34** Example – constructing the **Line** outline.

2. Call the **Extrude** tool and window the line outline.
3. Extrude to a height of **100**.

The stages of producing the extrusion are shown in Figs 12.34 and 12.35. The resulting 3D model is a surface model.

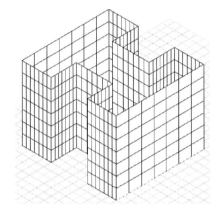

**Fig. 12.35** Example – **Dynamic Input.**

## NOTE →

The resulting 3D model shown in Fig. 12.35 is a surface model because the extrusion was constructed from an outline consisting of lines, which are individual objects in their own right. If the outline had been a polyline, the resulting 3D model would have been a solid model. The setting of **Mode** makes no difference.

## THE SWEEP TOOL

To call the tool *click* on its tool icon in the **Home/Create** panel (Fig. 12.36).

**Fig. 12.36** Selecting the **Sweep** tool from the **Home/Create** panel.

## EXAMPLE – **SWEEP** (FIG. 12.38)

**Fig. 12.37** Example **Sweep** – the outline to be swept.

1. Construct the pline outline in Fig. 12.37 in the **3D Navigation/ Top** view.
2. Change to the **3D Navigation/Front** view, **Zoom** to **1** and construct a pline as shown in Fig. 12.38 as a path central to the outline.
3. Make the layer **Magenta** current.
4. Place the window in the **3D Navigation/SW Isometric** view and *click* the **Sweep** tool icon.

The command line shows:

```
Command: _sweep
Current wire frame density: ISOLINES=4, Closed
 profiles creation mode = Solid
Select objects to sweep or [MOde]: _MO Closed
 profiles creation mode
[SOlid/SUrface] <Solid>: _SO
Select objects to sweep or [MOde]: pick the pline 1
 found
Select objects to sweep or [MOde]: right-click
Select sweep path or [Alignment/Base point/Scale/
 Twist]: pick the pline path
Command:
```

5. Place in **Visual Styles/Shaded**.

The result is shown in Fig. 12.38.

**Fig. 12.38** Example – **Sweep.**

## THE LOFT TOOL

To call the tool *click* on its icon in the **Home/Create** panel.

## EXAMPLE – LOFT (FIG. 12.41)

1. In the **3D Navigate/Top** view, construct the seven circles shown in Fig. 12.39 at vertical distances of **30** units apart.

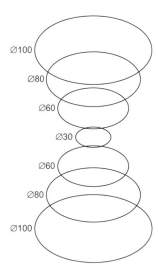

**Fig. 12.39** Example – **Loft** – the cross sections.

2. Place the drawing area in the **3D Navigate/SW Isometric** view.
3. Call the **Loft** tool with a *click* on its tool icon in the **Home/Create** panel (Fig. 12.40).

**Fig. 12.40** Selecting the **Loft** tool from the **Home/Create** panel.

4. Set **Cyan** as the current layer.

**5.** The command line shows:

```
Command: _loft
Select cross sections in lofting order or [POint/
 Join multiple curves]: pick 1 found
Select cross sections in lofting order or [POint/
 Join multiple curves]: pick 1 found, 2 total
Select cross sections in lofting order or [POint/
 Join multiple curves]: pick 1 found, 3 total
Select cross sections in lofting order or [POint/
 Join multiple curves]: pick 1 found, 4 total
Select cross sections in lofting order or [POint/
 Join multiple curves]: pick 1 found, 5 total
Select cross sections in lofting order or [POint/
 Join multiple curves]: pick 1 found, 6 total
Select cross sections in lofting order or [POint/
 Join multiple curves]: pick 1 found, 7 total
```
**Select cross sections in lofting order or [POint/
Join multiple curves]:** *enter* **j** *right-click*
```
Select curves that are to be joined into a single
 cross section: right-click 7 cross sections selected
Enter an option [Guides/Path/Cross sections only/
 Settings] <Cross sections only>: right-click
Command:
```

**6.** Place in **Visual Styles/Shaded with Edges**.

**Fig. 12.41** Example – **Loft**.

The result is shown in Fig. 12.41.

REVISION NOTES

1. In the AutoCAD 3D coordinate system, positive Z is towards the operator away from the monitor screen.
2. A 3D face is a mesh behind which other details can be hidden.
3. The Extrude tool can be used for extruding closed plines or regions to stated heights, to stated slopes, or along paths.
4. The Revolve tool can be used for constructing solids of revolution through any angle up to 360°.
5. 3D models can be constructed from Box, Sphere, Cylinder, Cone, Torus, and Wedge. Extrusions and/or solids of revolutions may form part of models constructed using these 3D tools.
6. The tools Union, Subtract, and Intersect are known as the Boolean operators.
7. When polylines forming an outline which is not closed are acted upon by the Extrude tool the resulting models will be 3D Surface models irrespective of the MOde setting.

## EXERCISES

Methods of constructing answers to the following exercises can be found in the free website:

**http://www.routledge.com/cw/yarwood**

The exercises which follow require the use of tools from the **Home/Create** panel in association with tools from other panels.

**1.** Fig. 12.42 shows the pline outline from which the polysolid outline in Fig. 12.43 has been constructed to a height of 100 and Width of 3. When the polysolid has been constructed, construct extrusions which can then be subtracted from the polysolid. Sizes of the extrusions are left to your judgment.

**Fig. 12.42** Exercise 1 – outline for polyline.

**Fig. 12.43** Exercise 1.

2. Fig. 12.44 shows a 3D model constructed from four polysolids which have been formed into a union using the **Union** tool from the **Home/Modify** panel. The original polysolid was formed from a hexagon of edge length **30**. The original polysolid was of height **40** and **Width 5**. Construct the union.

3. Fig. 12.45 shows the 3D model from Exercise 2 acted upon by the **Presspull** tool from the **Home/Create** panel. With the 3D model from Exercise 2 on screen and using the **Presspull** tool, construct the 3D model shown in Fig. 12.45. The distance of the pull can be estimated.

**Fig. 12.44** Exercise 2.

**Fig. 12.45** Exercise 3.

4. Construct the 3D model of a wine glass as shown in Fig. 12.47, working to the dimensions given in the outline drawing in Fig. 12.46.

   You will need to construct the outline and change it into a region before being able to change the outline into a solid of revolution using the **Revolve** tool from the **Home/Create** panel. This is because the semi-elliptical part of the outline has been constructed using the **Ellipse** tool, resulting in part of the outline being a spline, which cannot be acted upon by **Polyline Edit** to form a closed pline.

**Fig. 12.46** Exercise 4 – outline drawing.

**Fig. 12.47** Exercise 4.

**5.** Fig. 12.48 shows the outline from which a solid of revolution can be constructed. Use the **Revolve** tool from the **Home/Create** panel to construct the solid of revolution.

**Fig. 12.48** Exercise 5.

**6.** Construct a 3D solid model of a bracket working to the information given in Fig. 12.49.

**Fig. 12.49** Exercise 6.

**7.** Working to the dimensions given in Fig. 12.50 construct an extrusion of the plate to a height of 5 units.

**Fig. 12.50** Exercise 7.

**8.** Working to the details given in the orthographic projection in Fig. 12.51, construct a 3D model of the assembly. After constructing the pline outline(s) required for the solid(s) of revolution, use the **Revolve** tool to form the 3D solid.

**Fig. 12.51** Exercise 8.

**9.** Working to the polylines shown in Fig. 12.52 construct the **Sweep** shown in Fig. 12.53.

Profile outline                         Path

**Fig. 12.52** Exercise 9 – profile and path dimensions.

**Fig. 12.53** Exercise 9.

**10.** Construct the cross sections as shown in the left-hand drawing of Fig. 12.54 working to suitable dimensions. From the cross-sections construct the lofts shown in the right-hand view. The lofts are topped with a sphere constructed using the **Sphere** tool.

**Fig. 12.54** The cross-sections for Exercise 10.

# 13

# 3D MODELS IN VIEWPORTS

## AIM OF THIS CHAPTER

The aim of this chapter is to give examples of 3D solid models constructed in multiple viewport settings.

## THE 3D MODELING WORKSPACE

In Chapter 12 all 3D model actions were constructed in the **3D Basics** workspace. As shown in that chapter, a large number of different types of 3D models can be constructed in that workspace. In the following chapters 3D models will be constructed in the **3D Modeling** workspace, brought to screen with a *click* on from **3D Modeling** the **Workspace Switching** menu (Fig. 13.1). The AutoCAD window assumes the selected workspace settings (Fig. 13.2).

**Fig. 13.1** Opening the **3D Modeling** workspace.

If the **3D Modeling** workspace is compared with the **3D Basics** workspace (Fig.12.2 – page 218) it will be seen that there are several new tabs which when *clicked* bring changes in the ribbon with different sets of panels. In Fig. 13.2 the menu bar is included. This need not be included if the operator does not need the drop-down menus available from the menu bar.

**Introduction to AutoCAD 2012. ISBN: 978-0-08-096947-3**

**Fig. 13.2** The **3D Modeling** workspace in **SW Isometric** view and **Grid** on.

## SETTING UP VIEWPORT SYSTEMS

One of the better methods of constructing 3D models is in different multiple viewports. This allows what is being constructed to be seen from a variety of viewing positions. To set up multiple viewports please refer to the following directions.

In the **3D Modeling** workspace *click* **New** in the **View/Viewports** panel. From the popup list which appears (Fig. 13.3) select **Four: Equal**.

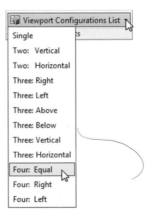

**Fig. 13.3** Selecting **Four: Equal** from the **View/Viewports** panel flyout.

The **Four: Equal** viewports layout appears (Fig. 13.4).

**Fig. 13.4** The **Four: Equal** viewports layout.

In Fig. 13.4 a simple 3D model has been constructed in the **Four: Equal** viewport layout. It will be seen that each viewport has a different view of the 3D model. Top right is an **isometric** view. Bottom right is a view from the **right** of the model. Bottom left is a view from the **left** of the model. Top left is a view from the **top** of the model. Note that the **front view** viewport is surrounded by a thicker line than the other three. The thicker line means that it is the **current** viewport. Any one of the four viewports can be made current with a *left-click* within its boundary. Note also that three of the views are in **third angle** projection.

When a viewport system has been opened it will usually be necessary to make each viewport current in turn. Use **Zoom** and **Pan** to ensure that all the views fit well within their boundaries.

If a **first angle** layout is needed, it will be necessary to open the **Viewports** dialog (Fig. 13.5).

This can be achieved with a *click* on the **New** icon in the **View/ Viewports** panel (Fig. 13.6).

First select **Four: Equal** from the **Standard viewports** list; select **3D** from the **Setup** popup menu; *click* in the **top right** viewport and select **Left** in the **Change View** popup list; and enter **first angle** in the **New name** field. Change the other viewports as shown. Save

**Fig. 13.5** The **Viewports** dialog set for a **3D first angle Four: Equal** setting.

**Fig. 13.6** Selecting **New** from the **View/Viewports** from the panel.

the settings with a *click* on the **Named Viewports** tab and *enter* the required name for the setup in the sub dialog which appears.

## FIRST EXAMPLE – FOUR: EQUAL VIEWPORTS (FIG. 13.9)

Fig. 13.7 shows a two-view orthographic projection of a support. To construct a **Scale 1:1** third angle 3D model of the support in a **Four: Equal** viewport setting on a layer color **Blue**:

1. Open a **Four: Equal** viewport setting from the **New** popup list in the **View/Viewports** panel (Fig. 13.3).

Dimensions in millimeters                           DO NOT SCALE

HOLE Ø40                    R20

Holes Ø20

30

10                    30

160

160

30                    30    20

30     60

| Name: | Scale: | Date: | Title: |
| A. Student | 1.2 | 23.11.2005 | Support 45/D |

**Fig. 13.7** First example – orthographic projection of the support.

2. *Click* in each viewport in turn, making the selected viewport active, and **Zoom** to **1**.
3. Using the **Polyline** tool, construct the outline of the plan view of the plate of the support, including the holes in the **Top** viewport (Fig. 13.5). Note the views in the other viewports.
4. Call the **Extrude** tool from the **Home/Modeling** panel and extrude the plan outline and the circles to a height of **20**.
5. With **Subtract** from the **Home/Solid Editing** panel, subtract the holes from the plate (Fig. 13.8).

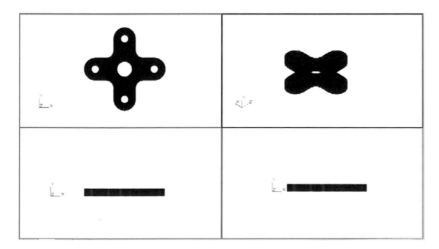

**Fig. 13.8** First example – the four viewports after Extrude and Subtract.

6. Call the **Box** tool and in the center of the plate construct a box of Width = **60**, Length = **60**, and Height = **30**.
7. Call the **Cylinder** tool and in the center of the box construct a cylinder of Radius = **20** and of Height = **30**.
8. Call **Subtract** and subtract the cylinder from the box.
9. *Click* in the **Right** viewport; with the **Move** tool, move the box and its hole into the correct position with regard to the plate.
10. With **Union**, form a union of the plate and box.
11. *Click* in the **Front** viewport and construct a triangle of one of the webs attached between the plate and the box. With **Extrude**, extrude the triangle to a height of **10**. With the **Mirror** tool, mirror the web to the other side of the box.
12. *Click* in the **Right** viewport and with the **Move** tool, move the two webs into their correct position between the box and plate. Then, with **Union**, form a union between the webs and the 3D model.
13. In the **Right** viewport, construct the other two webs and in the **Front** viewport, move, mirror, and union the webs as in steps **11** and **12**.

Fig. 13.9 shows the resulting 4-viewport scene.

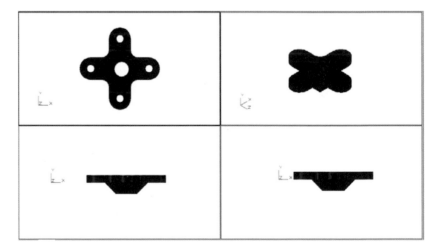

**Fig. 13.9** First example – **Four: Equal viewports.**

## SECOND EXAMPLE – FOUR: LEFT VIEWPORTS (FIG. 13.11)

1. Open a **Four: Left** viewport layout from the **Views/Viewports** popup list (Fig. 13.3).
2. Make a new layer of color **Magenta** and make that layer current.

3. In the **Top** viewport construct an outline of the web of the
   support bracket shown in Fig. 13.10. With the **Extrude** tool,
   extrude the parts of the web to a height of **20**.

**Fig. 13.10** Working drawing for the second example.

4. With the **Subtract** tool, subtract the holes from the web.
5. In the **Top** viewport, construct two cylinders central to the
   extrusion, one of radius **50** and height **30**, the second of radius
   **40** and height **30**. With the **Subtract** tool, subtract the smaller
   cylinder from the larger.
6. *Click* in the **Front** viewport and move the cylinders vertically by **5**
   units. With **Union** form a union between the cylinders and the web.
7. Still in the **Front** viewport and at one end of the union,
   construct two cylinders, the first of radius **10** and height **80**,
   the second of radius **15** and height **80**. Subtract the smaller
   from the larger.
8. With the **Mirror** tool, mirror the cylinders to the other end of
   the union.
9. Make the **Top** viewport current and with the **Move** tool, move
   the cylinders to their correct position at the ends of the union.
   Form a union between all parts on screen.
10. Make the **Isometric** viewport current. From the **View/Visual
    Styles** panel select **Conceptual**.

Fig. 13.11 shows the result.

**Fig. 13.11** Second example – **Four: Left** viewports.

## THIRD EXAMPLE – THREE: RIGHT VIEWPORTS (FIG. 13.13)

1. Open the **Three: Right** viewport layout from the **View/Viewports** popup list (Fig. 13.3).
2. Make a new layer of color **Green** and make that layer current.
3. In the **Front** viewport (top left-hand of screen), construct a pline outline to the dimensions in Fig. 13.12.

**Fig. 13.12** Third example – outline for solid of revolution.

4. Call the **Revolve** tool from the **Home/Modeling** panel and revolve the outline through 360°.
5. From the **View/Visual Styles** panel select **Conceptual**.

The result is shown in Fig. 13.13.

**Fig. 13.13** Third example – **Three: Right** viewports.

## NOTES →

1. When working in viewport layouts, make good use of the **Zoom** tool, because the viewports are smaller than a single viewport in AutoCAD 2012.
2. As in all other forms of constructing drawings in AutoCAD 2012 frequent toggling of **SNAP**, **ORTHO**, and **GRID** will allow speedier and more accurate working.

## REVISION NOTES↺

1. Outlines suitable for use when constructing 3D models can be constructed using the 2D tools such as Line, Arc, Circle, and polyline. Such outlines must be changed either to closed polylines or to regions before being incorporated in 3D models.
2. The use of multiple viewports can be of value when constructing 3D models in that various views of the model appear enabling the operator to check the accuracy of the 3D appearance throughout the construction period.

## EXERCISES

Methods of constructing answers to the following exercises can be found in the free website:

**http://www.routledge.com/cw/yarwood**

1. Using the Cylinder, **Box**, **Sphere**, and **Fillet** tools, together with the **Union** and **Subtract** tools and working to any sizes thought suitable, construct the "head" as shown in the **Three: Right** viewport setting as shown in Fig. 13.14.

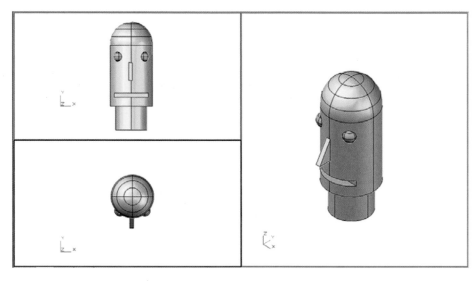

**Fig. 13.14** Exercise 1.

2. Using the tools **Sphere**, **Box**, **Union**, and **Subtract** and working to the dimensions given in Fig. 13.16, construct the 3D solid model as shown in the isometric drawing in Fig. 13.15.

**Fig. 13.15** Exercise 2.

**Fig. 13.16** Exercise 2 – working drawing.

3. Each link of the chain shown in Fig. 13.17 has been constructed using the tool **Extrude** and extruding a small circle along an elliptical path. Copies of the link were then made, half of which were rotated in a **Right** view and then moved into their position relative to the other links. Working to suitable sizes construct a link and from the link construct the chain as shown in Fig. 13.17.

**Fig. 13.17** Exercise 3.

4. A two-view orthographic projection of a rotatable lever from a machine is given in Fig. 13.18 together with an isometric drawing of the 3D model constructed to the details given in the drawing in Fig. 13.18.

**Fig. 13.18** Exercise 4 – orthographic projection.　　**Fig. 13.19** Exercise 4.

EXERCISES

5.  Construct the 3D model drawing (Fig. 13.19) in a **Four: Equal** viewport setting.

6.  Working in a **Three: Left** viewport setting, construct a 3D model of the faceplate to the dimensions given in Fig. 13.20. With the **Mirror** tool, mirror the model to obtain an opposite facing model. In the **Isometric** viewport call the **Hide** tool (Fig. 13.21).

**Fig. 13.20** Exercise 6 – dimensions.

**Fig. 13.21** Isometric view – Exercise 6.

# THE MODIFICATION OF 3D MODELS

## AIMS OF THIS CHAPTER

The aims of the chapter are:

**1.** To demonstrate how 3D models can be saved as blocks for insertion into other drawings via the **DesignCenter**.

**2.** To show how a library of 3D models in the form of blocks can be constructed to enable the models to be inserted into other drawings.

**3.** To give examples of the use of the tools from the **Home/Modify** panel:

3D Array – Rectangular and Polar 3D arrays.

**3D Mirror.**

**3D Rotate.**

**4.** To give examples of the use of the **Helix** tool.

**5.** To give an example of construction involving **Dynamic Input**.

**6.** To show how to obtain different views of 3D models in 3D space using the **View/Views/3D Manager** and the **ViewCube**

**7.** To give simple examples of surfaces using **Extrude**.

## CREATING 3D MODEL LIBRARIES

In the same way as 2D drawings of parts such as electronics symbols, engineering parts, building symbols, and the like can be saved in a file as blocks and then opened into another drawing, which can be done by *dragging* the appropriate block drawing from the DesignCenter, so can 3D models.

Introduction to AutoCAD 2012. ISBN: 978-0-08-096947-3

# FIRST EXAMPLE – INSERTING 3D BLOCKS (FIG. 14.4)

1. Construct 3D models of the parts for a lathe milling wheel holder to details as given in Fig. 14.1, each on a layer of different colors.

**Fig. 14.1** The components of a lathe milling wheel holder.

2. Save each of the 3D models of the parts to file names (see labels in Fig. 14.1) as blocks using **Create** from the **Insert/Block** panel. Save all seven blocks and delete the drawings on screen. Save the drawing with its blocks to a suitable file name (**Fig01.dwg**).
3. Set up a **Four: Equal** viewports setting.
4. Open the **DesignCenter** from the **View/Palettes** panel (Fig. 14.2) or by pressing the **Ctrl** and **2** keys of the keyboard.

**Fig. 14.2** Calling the **DesignCenter** from the **View/Palettes** panel.

5. In the **DesignCenter** *click* the directory **Chap14**, followed by
   another *click* on **Fig01.dwg** and yet another *click* on **Blocks**.
   The saved blocks appear as icons in the right-hand area of the
   **DesignCenter**.

6. *Drag and drop* the blocks one by one into one of the viewports
   on screen. Fig. 14.3 shows the **Nut** block ready to be *dragged*
   into the **Right** viewport. As the blocks are *dropped* on screen,
   they will need moving into their correct positions in suitable
   viewports using the **Move** tool from the **Home/Modify** panel.

**Fig. 14.3** First example – **Inserting 3D blocks**.

7. Using the **Move** tool, move the individual 3D models into their
   final places on screen and shade the **Isometric** viewport using
   **Conceptual** shading from the **Home/View** panel (Fig. 14.4).

**Fig. 14.4** First example – **Inserting 3D blocks**.

NOTES →

1. It does not matter which of the four viewports any one of the blocks is *dragged and dropped* into. The part automatically assumes the view of the viewport.

2. If a block destined for layer **0** is *dragged and dropped* into the layer **Centre** (which in our **acadiso.dwt** is of color **red** and of linetype **CENTER2**), the block will take on the color (**red**) and linetype of that layer (**CENTER2**).

3. In this example, the blocks are 3D models and there is no need to use the **Explode** tool option.

4. The examples of a **Four Equal** viewports screen shown in Figs 14.3 and 14.4 are in **first** angle. The front view is top-right; the end view is top-left; the plan is bottom right.

## SECOND EXAMPLE – A LIBRARY OF FASTENINGS (FIG. 14.6)

**Fig. 14.5** Second example – the five fastenings.

1. Construct 3D models of a number of engineering fastenings. In this example only five have been constructed – a 10 mm round head rivet, a 20 mm countersunk head rivet, a 20 mm cheese head bolt, a 20 mm countersunk head bolt, and a 20 mm hexagonal head bolt together with its nut (Fig. 14.5). With the **Create** tool save each separately as a block, erase the original drawings and save the file to a suitable file name – in this example Fig05.dwg.

2. Open the DesignCenter, and *click* on the **Chapter 14** directory, followed by a *click* on **Fig05.dwg.** Then *click* again on **Blocks** in the content list of **Fig05.dwg.** The five 3D models of fastenings appear as icons in the right-hand side of the DesignCenter (Fig. 14.6).

**Fig. 14.6** Second example – a library of fastenings.

**3.** Such blocks of 3D models can be *dragged and dropped* into position in any engineering drawing where the fastenings are to be included.

## CONSTRUCTING A 3D MODEL (FIG. 14.9)

A three-view projection of a pressure head is shown in Fig. 14.7.

**Fig. 14.7** Orthographic drawing for the example of constructing a 3D model.

To construct a 3D model of the head:

**1.** Select **Front** from the **View/Views** panel.
**2.** Construct the outline to be formed into a solid of revolution (Fig. 14.8) on a layer color **magenta** and with the **Revolve** tool, produce the 3D model of the outline.
**3.** Set the **View/Views/Top** view and with the **Cylinder** tool, construct cylinders as follows:

In the center of the solid – radius **50** and height **50**.

With the same center – radius **40** and height **40**. Subtract this cylinder from that of radius **50**.

At the correct center – radius **10** and height **25**.

At the same center – radius **5** and height **25**. Subtract this cylinder from that of radius **10**.

**4.** With the **Array** tool, form a polar **6** times array of the last two cylinders based on the center of the 3D model.

**Fig. 14.8** Example of constructing a 3D model – outline for solid of revolution.

5. Set the **View/Views/Front** view.
6. With the **Move** tool, move the array and the other two cylinders to their correct positions relative to the solid of revolution so far formed.
7. With the **Union** tool form a union of the array and other two solids.
8. Set the **View/Views/Right** view.
9. Construct a cylinder of radius **30** and height **25** and another of radius **25** and height **60** central to the lower part of the 3D solid so far formed.
10. Set the **View/Views/Top** view and with the **Move** tool move the two cylinders into their correct position.
11. With **Union**, form a union between the radius **30** cylinder and the 3D model and with **Subtract**, subtract the radius **25** cylinder from the **3D** model.
12. *Click* **Realistic** in the **View/Visual Styles** panel list.

**Fig. 14.9** Example of constructing a 3D model.

The result is given in Fig. 14.9.

> NOTE →
>
> This 3D model could equally as well have been constructed in a three or four viewports setting. **Full Shading** has been set on from the **Render** ribbon, hence the line of shadows.

**Fig. 14.10** Example – **3D Array** – the star pline.

## THE 3D ARRAY TOOL

### FIRST EXAMPLE – A RECTANGULAR ARRAY (FIG. 14.12)

1. Construct the star-shaped pline on a layer color green (Fig. 14.10) and extrude it to a height of **20**.
2. *Click* on the **3D Array** in the **Home/Modify** panel (Fig. 14.11).

The command line shows:

```
Command: _3darray
Select objects: pick the extrusion 1 found
Select objects: right-click
Enter the type of array [Rectangular/Polar] <R>:
   right-click
Enter the number of rows (---) <1>: enter 3 right-
   click
Enter the number of columns (III): enter 3 right-click
Enter the number of levels (...): enter 4 right-click
Specify the distance between rows (—): enter 100
   right-click
```

**Fig. 14.11** Selecting **3D Array** from the **Modify** drop-down menu.

```
Specify the distance between columns (III): enter
  100 right-click
Specify the distance between levels (...): enter
  300 right-click
Command:
```

**3.** Place the screen in the **View/Views/SW Isometric** view.
**4.** Shade using the **View/Visual Styles/Shaded with Edges** visual style (Fig. 14.12).

## SECOND EXAMPLE – A POLAR ARRAY (FIG. 14.13)

**1.** Use the same star-shaped 3D model.
**2.** Call the **3D Array** tool again. The command line shows:

```
Command: _3darray
Select objects: pick the extrusion 1 found
Select objects: right-click
Enter the type of array [Rectangular/Polar] <R>:
  enter p (Polar) right-click
Enter number of items in the array: 12
Specify the angle to fill (+=ccw), -=cw) <360>:
  right-click
Rotate arrayed objects? [Yes/No] <Y>: right-click
```

**Fig. 14.12** First example – a **3D Array**.

**Fig. 14.13** Second example – a **3D Polar Array**.

**Fig. 14.14** Third example – a **3D Polar Array** – the 3D model to be arrayed.

```
Specify center point of array: 235,125
Specify second point on axis of rotation: 300,200
Command:
```

3. Place the screen in the **View/Views/SW Isometric** view.
4. Shade using the **View/Visual Styles Shaded** visual style (Fig. 14.13).

## THIRD EXAMPLE – A POLAR ARRAY (FIG. 14.15)

1. Working on a layer of color **red**, construct a solid of revolution in the form of an arrow to the dimensions as shown in Fig. 14.14.
2. *Click* **3D Array** in the **Home/Modify** panel. The command line shows:

```
Command: _3darray
Select objects: pick the arrow 1 found
Select objects: right-click
Enter the type of array [Rectangular/Polar]<R>:
    enter p right-click
Enter the number of items in the array: enter 12
    right-click
Specify the angle to fill (+=ccw, -=cw) <360>: right-
    click
Rotate arrayed objects? [Yes/No] <Y>: right-click
Specify center point of array: enter 40,170,20
    right-click
Specify second point on axis of rotation: enter
    60,200,100 right-click
Command:
```

3. Place the array in the **3D Navigate/SW Isometric** view and shade to **View/Visual Styles/Shades of Gray**. The result is shown in Fig. 14.15.

**Fig. 14.15** Third example – a **3D Polar Array**.

# THE 3D MIRROR TOOL

## FIRST EXAMPLE – 3D MIRROR (FIG. 14.17)

**Fig. 14.16** First example –
**3D Mirror** – outline of object
to be mirrored.

1. Working on a layer of color **green,** construct the outline in Fig. 14.16.
2. Extrude the outline to a height of **20.**
3. Extrude the region to a height of **5** and render. A **Conceptual** style shading is shown in Fig. 14.17 (left-hand drawing).
4. *Click* on **3D Mirror** in the **3D Operation** sub-menu of the **Modify** drop-down menu. The command line shows:

```
Command: _3dmirror
Select objects: pick the extrusion 1 found
Select objects: right-click
Specify first point of mirror plane (3 points): pick
Specify second point on mirror plane: pick
Specify third point on mirror plane or [Object/Last/
   Zaxis/View/XY/YZ/ZX/3points]: enter .xy right-
   click
of (need Z): enter 1 right-click
Delete source objects? [Yes/No]: <N>: right-click
Command:
```

The result is shown in the right-hand illustration of Fig. 14.17.

**Fig. 14.17** First example – **3D Mirror** – before and after **Mirror.**

# SECOND EXAMPLE – 3D MIRROR (FIG. 14.19)

1. Construct a solid of revolution in the shape of a bowl in the **3D Navigate/Front** view working on a layer of color **magenta** (Fig. 14.18).

**Fig. 14.18** Second example **3D Mirror** – the 3D model.

2. *Click* **3D Mirror** in the **Home/Modify/**panel. The command line shows:

```
Command: _3dmirror
Select objects: pick the bowl 1 found
Select objects: right-click
Specify first point on mirror plane (3 points): pick
Specify second point on mirror plane: pick
Specify third point on mirror plane: enter .xy
  right-click
(need Z): enter 1 right-click
Delete source objects:? [Yes/No]: <N>: right-click
Command:
```

The result is shown in Fig. 14.19.

3. Place in the **3D Navigate/SW Isometric** view.
4. Shade using the **View/Visual Styles Conceptual** visual style (Fig. 14.19).

**Fig. 14.19** Second example – **3D Mirror** – the result in a front view.

## THE 3D ROTATE TOOL

### EXAMPLE – 3D ROTATE (FIG. 14.20)

1. Use the same 3D model of a bowl as for the last example. *Pick* **3D Rotate** tool from the **Home/Modify** panel. The command line shows:

```
Command: _3drotate
Current positive angle in UCS:
  ANGDIR=counterclockwise ANGBASE=0
Select objects: pick the bowl 1 found
Select objects: right-click
Specify base point: pick the center bottom of the
  bowl
Specify rotation angle or [Copy/Reference] <0>:
  enter 60 right-click
Command
```

2. Place in the **3D Navigate/SW Isometric** view and in **Conceptual** shading.

**Fig. 14.20** Example – **3D Rotate**.

## THE SLICE TOOL

### FIRST EXAMPLE – SLICE (FIG. 14.24)

1. Construct a 3D model of the rod link device shown in the two-view projection in Fig. 14.21 on a layer of color **green**.

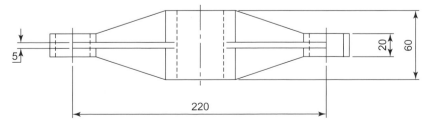

**Fig. 14.21** First example – **Slice** – the two-view drawing.

2. Place the 3D model in the **3D Navigation/Top** view.
3. *Call* the **Slice** tool from the **Home/Solid Editing** panel (Fig. 14.22).

**Fig. 14.22** The **Slice** tool icon from the **Home/Solid Editing** panel.

The command line shows:

```
Command: _slice
Select objects: pick the 3D model
Select objects to slice: right-click
Specify start point of slicing plane or [planar
   Object/Surface/Zaxis/View/XY/YZ/ZX/3points]
   <3points>: pick
Specify second point on plane: pick
Specify a point on desired side or [keep Both sides]
   <Both>: right-click
Command:
```

Fig. 14.23 shows the *picked* points.

**Fig. 14.23** First example – **Slice** – the *picked* points.

4. With the **Move** tool, move the lower half of the sliced model away from the upper half.
5. Place the 3D model(s) in the **ViewCube/Isometric** view.
6. Shade in **Conceptual** visual style. The result is shown in Fig. 14.24.

## SECOND EXAMPLE – SLICE (FIG. 14.25)

1. On a layer of color **green**, construct the closed pline shown in the left-hand drawing Fig. 14.25 and with the **Revolve** tool, form a solid of revolution from the pline.

**Fig. 14.24** First example – **Slice**.

2. With the **Slice** tool and working to the same sequence as for the first **Slice** example, form two halves of the 3D model.
3. Place in **View/Views/Visual Styles/X-Ray**.
The right-hand illustration of Fig. 14.25 shows the result.
4. Place the model in the **3D Navigate/Front** view, **Zoom** to **1** and **Move** its parts apart.
5. Make a new layer **Hatch** of color **magenta** and make the layer current.

**Fig. 14.25** Second example – **Slice**.

## VIEWS OF 3D MODELS

Some of the possible viewing positions of a 3D model which can be obtained by using the **View/Views 3D Navigation** popup list have already been shown in earlier pages. Fig. 14.27 shows the viewing positions of the 3D model of the arrow (Fig. 14.26) using the viewing positions from the **3D Navigation** popup.

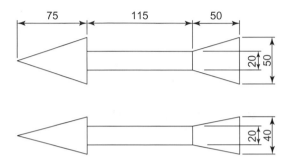

**Fig. 14.26** Two views of the arrowFig.

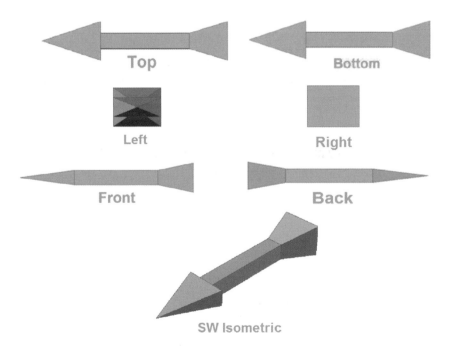

**Fig. 14.27** Views using the **View/Views 3D Navigation** popup list.

## THE VIEWCUBE

Another method of obtaining viewing positions of a 3D model is by using the **ViewCube**, which can usually be seen at the top-right corner of the AutoCAD 2012 window (Fig. 14.28). The **ViewCube**

can be turned off by *entering* **navvcubedisplay** at the command line
and *entering* **1** as a response as follows:

```
Command: navvcubedisplay
Enter new value for NAVVCUBEDISPLAY <3>: enter 1
  right-click
```

*Entering* **3** as a response to **navvcubedisplay** causes
the **ViewCube** to reappear.

The **ViewCube** is used as follows:

*Click* on **Top** and the **Top** view of a 3D model appears.

*Click* on **Front** and the **Front** view of a 3D model
appears.

And so on. *Clicking* the arrows at top, bottom or
sides of the **ViewCube** moves a model between
views.

A *click* on the house icon at the top of the **ViewCube**
places a model in the **SW Isometric** view.

Fig. 14.28 The ViewCube.

## USING DYNAMIC INPUT TO CONSTRUCT A HELIX

As with all other tools (commands) in AutoCAD 2012 a helix
can be formed working with the **Dynamic Input** (DYN) system.
Fig. 14.30 shows the stages (**1** to **5**) in the construction of the helix
in the second example.

Set **DYN** on with a *click* on its button in the status bar.

1. *Click* the **Helix** tool icon in the **Home/Draw** panel (Fig. 14.29).
   The first of the **DYN** prompts appears.

Fig. 14.29 The Helix tool in the Home/Draw panel.

*Enter* the following at the command line using the **down** kcy of the
keyboard when necessary.

```
Command: _Helix
Number of turns = 10 Twist=CCW
Specify center point of base: enter 95,210
Specify base radius or [Diameter]: enter 55
Specify top radius or [Diameter]: enter 35
Specify helix height or [Axis endpoint/Turns/turn
    Height/tWist]: enter 100
Command:
```

Fig. 14.30 shows the sequence of **DYN** tooltips and the completed helix.

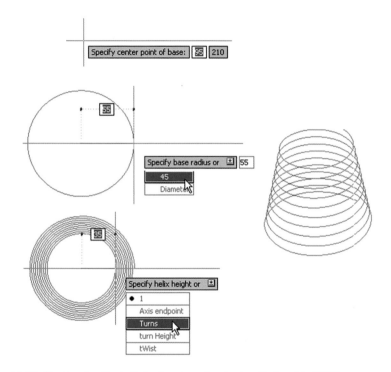

**Fig. 14.30** Constructing the helix for the second example with the aid of **DYN**.

## 3D SURFACES

As mentioned on page 224, surfaces can be formed using the **Extrude** tool on lines and polylines. Two examples are given below in Figs 14.32 and 14.44.

## FIRST EXAMPLE – 3D SURFACE (FIG. 14.32)

**1.** In the **ViewCube/Top** view, on a layer color **magenta**, construct the polyline in Fig. 14.31.

**Fig. 14.31** First example – **3D Surface** – polyline to be extruded.

2. In the **ViewCube/Isometric** view, call the **Extrude** tool from the **Home/Modeling** control and extrude the polyline to a height of 80. The result is shown in Fig. 14.32.

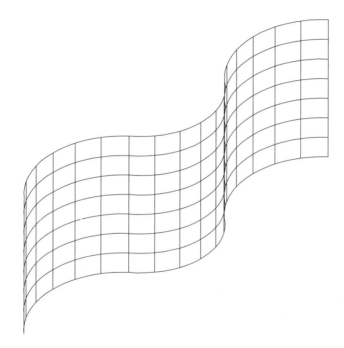

**Fig. 14.32** First example – **3D Surface**.

## SECOND EXAMPLE – 3D SURFACE (FIG. 14.34)

1. In the **Top** view on a layer of color **blue** construct the circle in Fig. 14.33 using the **Break** tool and break the circle as shown.

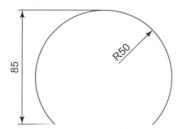

**Fig. 14.33** Second example – **3D Surface**. The part-circle to be extruded.

**Fig. 14.34** Second example –
**3D Surface**.

2. In the **3D Manager/SW Isometric** view, call the **Extrude** tool and extrude the part-circle to a height of **80**. Shade in the **Conceptual** visual style (Fig. 14.34).

The result is shown in Fig. 14.34

### REVISION NOTES

1. 3D models can be saved as blocks in a similar manner to the method of saving 2D drawings as blocks.
2. Libraries can be made up from 3D model drawings.
3. 3D models saved as blocks can be inserted into other drawings via the DesignCenter.
4. Arrays of 3D model drawings can be constructed in 3D space using the 3D Array tool.
5. 3D models can be mirrored in 3D space using the 3D Mirror tool.
6. 3D models can be rotated in 3D space using the 3D Rotate tool.
7. 3D models can be cut into parts with the Slice tool.
8. Helices can be constructed using the Helix tool.
9. Both the View/View/Navigation popup list and the ViewCube can be used for placing 3D models in different viewing positions in 3D space.
10. The Dynamic Input (DYN) method of construction can be used equally as well when constructing 3D model drawings as when constructing 2D drawings.
11. 3D surfaces can be formed from polylines or lines with Extrude.

## EXERCISES

Methods of constructing answers to the following exercises can be found in the free website:

**http://www.routledge.com/cw/yarwood**

1. Fig. 14.35 shows a **Realistic** shaded view of the 3D model for this exercise. Fig. 14.36 gives a three-view projection of the model. Working to the details given in Fig. 14.36, construct the 3D model.

**Fig. 14.35** Exercise 1.

**Fig. 14.36** Exercise 1 – a three-view projection.

2. Construct a 3D model drawing of the separating link shown in the two-view projection (Fig. 14.37). With the **Slice** tool, slice the model into two parts and remove the rear part. Place the front half in an isometric view using the **ViewCube** and shade the resulting model.

**Fig. 14.37** Exercise 2.

3. Working to the dimensions given in the two orthographic projections in Fig. 14.38, and working on two layers of different colors, construct an assembled 3D model of the one part inside the other. With the **Slice** tool, slice the resulting 3D model into two equal parts, and place in an isometric view. Shade the resulting model in **Realistic** mode as shown in Fig. 14.39.

**Fig. 14.38** Exercise 3 – orthographic projection.

**Fig. 14.39** Exercise 3 – orthographic projections.

4. Construct a solid of revolution of the jug shown in the orthographic projection in Fig. 14.40. Construct a handle from an extrusion of a circle along a semicircular path. **Union** the two parts. Place the 3D model in a suitable isometric view and render.

Torus R67.5 tube R2.5

Cylinder R70 and 250 high

Cylinder R65 and 245 high

R58

R72

Torus R50 tube R7

Torus R70 tube R5

250

245

**Fig. 14.40** Exercise 4.

**5.** In the **Top** view on a layer color **blue** construct the four polylines in Fig. 14.41. Call the **Extrude** tool and extrude the polylines to a height of **80** and place in the **Isometric** view. Then call **Visual Styles/Shades of Gray** shading (Fig. 14.42).

**Fig. 14.41** Exercise 5 – outline to be extruded.

**Fig. 14.42** Exercise 5.

**6.** In **3D Navigation/Right** view construct the lines and arc in Fig. 14.43 on a layer color **green**. Extrude the lines and arc to a height of **180**; place in the **SW Isometric** view and in the shade style **Visual Styles/Realistic** (Fig. 14.44).

**Fig. 14.43** Exercise 6 – outline to be extruded.

**Fig. 14.44** Exercise 6.

# RENDERING

## AIMS OF THIS CHAPTER

The aims of this chapter are:

1. To construct a template for 3D Modeling to be used as the drawing window for further work in 3D in this book.
2. To introduce the use of the **Render** tools in producing photographic like images of 3D solid models.
3. To show how to illuminate a 3D solid model to obtain good lighting effects when rendering.
4. To give examples of the rendering of 3D solid models.
5. To introduce the idea of adding materials to 3D solid models in order to obtain a realistic appearance to a rendering.
6. To demonstrate the use of the forms of shading available using **Visual Styles** shading.
7. To demonstrate methods of printing rendered 3D solid models.
8. To give an example of the use of a camera.

## SETTING UP A NEW 3D TEMPLATE

In this chapter we will be constructing all 3D model drawings in the **3dacadiso. dwt** template. The template is based on the **3D Modeling** workspace shown on page 250 in Chapter 13.

1. *Click* the **Workspace Switching** button and *click* **3D Modeling** from the menu (Fig. 15.1).

Fig. 15.1  *Click* **3D Modeling** in the **Workspace Settings** menu.

**2.** The AutoCAD window will appear (Fig. 15.2).

Fig. 15.2  The **3D Modeling** workspace.

**3.** Set **Units** to a **Precision** of **0**, **Snap** to **5**, and **Grid** to **10**. Set **Limits** to **420,297**. **Zoom** to **All**.

**4.** In the **Options** dialog *click* the **Files** tab and *click* **Default Template File Name for QNEW** followed by a *double-click* on the file name which appears. This brings up the **Select Template** dialog, from which the **3dacadiso.dwt** can be selected. Now when AutoCAD 2012 is opened from the Windows desktop, the 3dacadiso.dwt template will open.

5. Set up five layers of different colors named after the colors.
6. Save the template to the name **3dacadiso** and then *enter* a suitable description in the **Template Definition** dialog.

## THE MATERIALS BROWSER PALETTE

*Click* **Materials Browser** in The **Render/Materials** palette (Fig. 15.3). The **Materials Browser** palette appears *docked* at an edge of the AutoCAD window. *Drag* the palette away from its *docked* position. *Click* the arrow to the left of **Autodesk Library** and in the list which appears, *click* **Brick**. A list of brick icons appears in a list to the right of the **Autodesk Library** list (Fig. 15.4).

**Fig. 15.3** The **Materials Browser** button in the **Render/Materials** panel.

The **Materials Browser** palette can be *docked* against either side of the AutoCAD window if needed.

## APPLYING MATERIALS TO A MODEL

Materials can be applied to a 3D model from selection of the icons in the **Materials Browser** palette. Three examples follow – applying a **Brick** material, applying a **Metal** material, and applying a **Wood** material.

In the following three examples lighting effects are obtained by *clicking* the **Sun Status** icon in the **Render/Sun & Location** panel (Fig. 15.5).

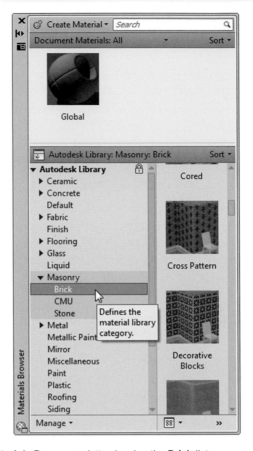

**Fig. 15.4** The **Materials Browser** palette showing the **Brick** list.

**Fig. 15.5** The **Sun Status** button in the **Render/Sun & Location** panel.

The command line shows:

```
Command: _sunstatus
Enter new value for SUNSTATUS <1>: 0
Command:
```

When the material has been applied, *click* **Render Region** from the sub-panel of the **Render/Render** panel (Fig. 15.6) and after selecting a window surrounding the model, the model renders.

**Fig. 15.6** The **Render Region** button from the **Render/Render** panel.

## FIRST EXAMPLE – APPLYING A BRICK MATERIAL (FIG. 15.7)

Construct the necessary 3D model (Fig. 15.7). In the **Materials Browser** palette, in the **Autodesk Library** list *click* **Brick**. A number of icons appear in the right-hand column of the palette representing different brick types. *Pick* the **Brown Modular** icon from the list. The icon appears in the **Materials in this document** area of the palette. *Right-click* in the icon and in the menu which appears select **Assign to Selection**. *Click* the model. Select **Render Region** from the **Render/Render** panel (Fig. 15.6). Window the model. The model renders (Fig. 15.7).

**Fig. 15.7** First example – **assigning a Masonry Brick material**.

# SECOND EXAMPLE – APPLYING A METAL MATERIAL (FIG. 15.8)

Construct the necessary 3D model. From the **Materials Browser** palette *click* **Metals** in the **Autodesk Library** list. Select **polished Brass 7** from the metal icons. *Click* **Assign to Selection** from the *right-click* menu in the **Materials in this document** area and *click* the model. Then with the **Render Region** tool render the model (Fig. 15.8)

**Fig. 15.8** Second example – **assigning a Metal material.**

# THIRD EXAMPLE – APPLYING A WOOD MATERIAL (FIG. 15.9)

Construct the necessary 3D model – a board. In the **Materials Browser** palette *click* **Wood** In the **Autodesk Library** list. Select **Pine Coarse** from the metal icons. *Click* **Assign to Selection** from the *right-click* menu in the **Materials in this document** area and *click* the model. Then with the **Render Region** tool render the model (Fig. 15.9).

# MODIFYING AN APPLIED MATERIAL

If the result of applying a material direct to a model from the selected materials palette is not satisfactory, modifications to the

**Fig. 15.9** Third example – **assigning a Wood material.**

applied material can be made. In the case of the third example, *double-click* on the chosen material icon in the **Materials Browser** palette and the **Materials Editor** palette will appear showing the materials in the drawing (Fig. 15.10). Different features, such as changing the color of the applied material or choosing a different texture, are possible.

**Fig. 15.10** The **Materials Browser** palette showing the materials in a 3D model and the **Materials Editor Open File** dialog.

## The Materials Editor Palette

1. A *click* in the **Image** area of the palette brings the **Materials Editor Open File** dialog to screen. From this dialog a very large number of material images can be chosen. The images are kept in the folders:

**C:\Program Files\Common Files\Autodesk\Shared\Materials 2001\ asset library fbm.\1\Mats (or 2\Mats or 3\Mats).**

2. A *click* in the **Color** field brings the **Select Color** dialog to screen. This is where a color can be selected for the material and changed.

3. *Click* in the check boxes named **Reflectivity, Transparency** etc. to bring up features which can amend the material being edited.

Experimenting with this variety of settings in the **Materials Editor** palette allows amending the material to the operator's satisfaction.

---

NOTE →

Material bitmaps are kept in the folders.

**C:\Program Files\Common Files\Autodesk\Shared\Materials 2001\asset library fbm.\1\Mats (or 2\Mats or 3\Mats).**

---

## FOURTH EXAMPLE – AVAILABLE MATERIALS IN DRAWING (FIG. 15.11)

As an example Fig. 15.11 shows the five of the materials applied to various parts of a 3D model of a hut in a set of fields surrounded by fences. The **Materials Browser** is shown. A *click* on a material in the **Available Materials in Drawing** brings the **Materials Editor** palette to screen, in which changes can be made to the selected material.

## THE RENDER TOOLS AND DIALOGS

The tool icons and menus in the **Render/Render** sub-panel are shown in Fig. 15.12.

A *click* in the outward facing arrow at the bottom right-hand corner of the **Render/Render** panel brings the **Advanced Render Settings** palette on screen. Note that a *click* on this arrow if it appears in any panel will bring either a palette or a dialog on screen.

**Fig. 15.11** An example of materials applied to parts of a 3D model.

**Fig. 15.12** The tools and menus in the **Render/Render** panel.

## THE LIGHTS TOOLS

The different forms of lighting from light palettes are shown in Fig. 15.13.

There are a large number of different types of lighting available when using AutoCAD 2012, among which those most frequently used are:

Default lighting: Depends on the setting of the set variable.
Point: lights shed light in all directions from the position in which the light is placed.

**Fig. 15.13** **Lighting** buttons and menus in the **Render/Lights** panel.

**Distant:** lights send parallel rays of light from their position in the direction chosen by the operator.

**Spotlights:** illuminate as if from a spotlight. The light is in a direction set by the operator and is in the form of a cone, with a "hotspot" cone giving a brighter spot on the model being lit.

**Sun:** light that can be edited to position.

**Sky background and illumination.**

A variety of lights of different types in which lights of a selected wattage which can be placed in a lighting scene are available from the **Tool Palettes – All Palettes** palette. These are shown in Fig. 15.14.

The set variable **LIGHTINGUNITS** must be set to **1** or **2** for these lights to function. To set this variable:

```
Command: enter lightingunits right-click
Enter new value for LIGHTINGUNITS <2>:
```

Settings are:

**0:** No lighting units are used and standard (generic) lighting is enabled.

**1:** American lighting units (foot-candles) are used and photometric lighting is enabled.

**2:** International lighting units (lux) are used and photometric lighting is enabled.

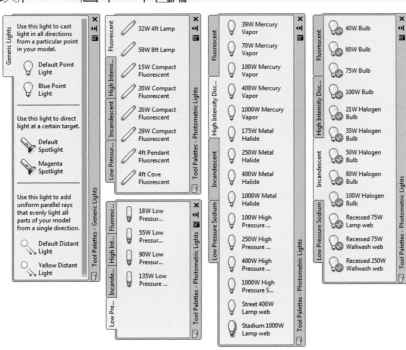

**Fig. 15.14** The **Lighting** tool palettes.

**Note:** In the previous examples of rendering, **Generic lighting** was chosen.

## Placing Lights to Illuminate a 3D Model

In this book examples of lighting methods shown will only be concerned with the use of **Point, Direct,** and **Spot** lights, together with **Default lighting,** except for the example given on page 310, associated with using a camera.

Any number of the three types of lights – **Point, Distant,** and **Spotlight,** can be positioned in 3D space as wished by the operator.

In general, good lighting effects can be obtained by placing a **Point** light high above the object(s) being illuminated, a **Distant** light pointing toward the object at a distance from the front and above the general height of the object(s), and a second **Distant** light pointing toward the object(s) from one side and not as high as the other **Distant** light. If desired, **Spotlights** can be used either on their own or in conjunction with the other two forms of lighting.

## SETTING RENDERING BACKGROUND COLOR

The default background color for rendering in the 3dacadiso template is black by default. In this book, all renderings are shown on a white background in the viewport in which the 3D model drawing was constructed. To set the background to white for renderings:

**1.** At the command line:

Command: *enter* view *right-click*

The **View Manager** dialog appears (Fig. 15.15). *Click* **Model View** in its **Views** list, followed by a *click* on the **New...** button.

**Fig. 15.15** The **View Manager** dialog.

**2.** The **New View/Shot Properties** dialog (Fig. 15.16) appears. *Enter* **current** (or similar) in the **View name** field. In the **Background** popup list *click* **Solid**.

The **Background** dialog appears (Fig. 15.17).

**3.** In the **Background** dialog *click* in the **Color** field. The **Select Color** dialog appears (Fig. 15.17).

**4.** In the **Select Color** dialog *drag* the slider as far upwards as possible to change the color to white (**255,255,255**). Then *click* the dialog's **OK** button. The **Background** dialog reappears showing white in the **Color** and **Preview** fields. *Click* the **Background** dialog's **OK** button.

**5.** The **New View/Shot Properties** dialog reappears showing **current** highlighted in the **Views** list *Click* the dialog's **OK** button.

**6.** The **View Manager** dialog reappears. *Click* the **Set Current** button, followed by a *click* on the dialog's **OK** button (Fig. 15.18).

**7.** *Enter* **rpref** at the command line. The **Advanced Render Settings** palette appears. In the palette, in the **Render Context** field *click*

Fig. 15.16 The **New View/Shot Properties** dialog.

Fig. 15.17 The **Background** dialog.

the arrow to the right of **Window** and in the popup menu which appears *click* **Viewport** as the rendering destination (Fig. 15.19).

8. Close the palette and save the screen with the new settings as the template **3dacadiso.dwt**. This will ensure renderings are made in the workspace in which the 3D model was constructed to be the same workspace in which renderings are made – on a white background.

**Fig. 15.19** The **Advanced Render Settings** dialog.

**Fig. 15.18** The **View Manager** dialog.

## FIRST EXAMPLE – RENDERING (FIG. 15.28)

1. Construct a 3D model of the wing nut shown in the two-view projection in Fig. 15.20.
2. Place the 3D model in the **3D Navigation/Top** view, **Zoom** to **1** and with the **Move** tool, move the model to the upper part of the AutoCAD drawing area.
3. *Click* the **Point Light** tool icon in the **Render/Lights** panel (Fig. 15.21).

**Fig. 15.20** First example – **Rendering** – two-view projection.

**Fig. 15.21** The **Point Light** icon in the **Render/Lights** panel.

The warning window appears (Fig.15.22). *Click* **Turn off** the **default lighting** in the window.

**Fig. 15.22** The **Lighting – Viewport Lighting Mode** warning window.

4. A **New Point Light** icon appears (depending upon the setting of the **Light Glyph Setting** in the **Drafting** area of the **Options** dialog) and the command line shows:

```
Command: _pointlight
Specify source location <0,0,0>: enter .xy right-
 click of pick center of model (need Z): enter 500
 right-click
Enter an option to change [Name/Intensity/Status/
 shadoW/Attenuation/Color/eXit]
<eXit>: enter n right-click
Enter light name <Pointlight1>: enter Point01 right-
 click
Enter an option to change [Name/Intensity/Status/
 shadoW/Attenuation/Color/eXit]<eXit>: right-click
Command:
```

5. There are several methods by which **Distant** lights can be called: by selecting **Default Distant Light** from the **Generic Lights** palette, with a *click* on the **Distant** icon in the **Render/Lights** panel, or by *entering* **distantlight** at the command line.

No matter which method is adopted the **Lighting – Viewport Lighting Mode** dialog (Fig. 15.22) appears. *Click* **Turn off** the **default lighting (recommended)**. The **Lighting – Photometric Distant Lights then** dialog appears (Fig. 15.23).

*Click* **Allow distant lights** in this dialog and the command line shows:

```
Command: _distantlight
Specify light direction FROM <0,0,0> or [Vector]:
 enter .xy right-click
of pick a point below and to the left of the model
 (need Z): enter 400 right-click
```

**Fig. 15.23** The **Photometric Distant Lights** dialog.

```
Specify light direction TO <1,1,1>: enter .xy right-
   click
of pick a point at the center of the model (need Z):
   enter 70 right-click
Enter an option to change [Name/Intensity/Status/
   shadoW/Color/eXit] <eXit>: enter n right-click
Enter light name <Distantlight8>: enter Distant01
   right-click
Enter an option to change [Name/Intensity/Status/
   shadoW/Color/eXit] <eXit>: right-click
Command:
```

**Fig. 15.24** The arrow at the bottom of the **Render/Lights** panel.

6. Place another **Distant Light** (**Distant2**) in front of and below the model **FROM Z** of **300** and at the same position **TO the model.**
7. When the model has been rendered, if a light requires to be changed in intensity, shadow, position or color, *click* the arrow at the bottom right-hand corner of the **Render/Lights** panel (Fig. 15.24) and the **Lights in Model** palette appears (Fig. 15.25). *Double-click* a light name in the palette and the **Properties** palette for the elected light appears into which modifications can be made (Fig. 15.25). Amendments can be made as necessary.

NOTES →

1. In this example the **Intensity factor** has been set at **0.5** for lights. This is possible because the lights are close to the model. In larger size models the **Intensity factor** may have to be set to a higher figure.
2. Before setting the **Intensity factor** to **0.5**, **Units** need setting to **OO** in the **Drawing Units** dialog (See Chapter 1).

Fig. 15.25 The **Lights in Model** and **Properties** palettes.

## Assigning a Material to the Model

1. Open the **Materials Browser** palette with a *click* on the **Materials Browser** icon in the **Render/Materials** panel. From the **Autodesk Library** list in the palette, select **Metal**. When the icons for the metals appear in the right-hand column of the palette, *double-click* **Brass Polished**. The icon appears in the **Materials in this document** area of the palette (Fig. 15.26).

Fig. 15.26 The **Materials Browser** and the rendering.

2. *Click* **Assign to Selection** in the *right-click* menu of the material in the **Materials Browser** palette, followed by a *click* on the model, followed by *left-click* when the model has received the assignment.

3. Select **Presentation** from the **Render Presets** menu in the sub **Render/Render** panel (Fig. 15.27).

**Fig. 15.27** Setting the form of rendering to **Presentation**.

4. Render the model (Fig. 15.29) using the **Render Region** tool from the **Render/Render** panel and if now satisfied save to a suitable file name.

**Fig. 15.29** Second example – **Rendering**.

**Fig. 15.28** Second example – **Rendering** – orthographic projection.

NOTE →

The limited description of rendering given in these pages does not show the full value of different types of lights, materials, and rendering methods. The reader is advised to experiment further with the facilities available for rendering.

# SECOND EXAMPLE – RENDERING A 3D MODEL (FIG. 15.29)

1. Construct 3D models of the two parts of the stand and support given in the projections in Fig. 15.28 with the two parts assembled together.
2. Place the scene in the **ViewCube/Top** view, **Zoom** to **1**, and add lighting.
3. Add different materials to the parts of the assembly and render the result.

Fig. 15.29 shows the resulting rendering.

# THIRD EXAMPLE – RENDERING (FIG. 15.31)

Fig. 15.30 shows an exploded, rendered 3D model of a pumping device from a machine and Fig. 15.31 shows a third angle orthographic projection of the device.

**Fig. 15.30** Third example – **Rendering**.

## FREE ORBIT

# EXAMPLE – FREE ORBIT (FIG. 15.33)

Place the second example in a **Visual Styles/Conceptual** shading.

*Click* the **Free Orbit** button in the **View/Navigate** panel (Fig. 15.32).

An orbit cursor appears on screen. Moving the cursor under mouse control allows the model on screen to be placed in any desired viewing position. Fig. 15.33 shows an example of a **Free Orbit**.

*Right-click* anywhere on screen and a *right-click* menu appears.

**Fig. 15.31** Third example – **rendering** – **exploded orthographic views**.

**Fig. 15.32** The **Free Orbit** tool from the **View/Navigate** panel.

**Fig. 15.33** Example – **Free Orbit**.

## PRODUCING HARDCOPY

Printing or plotting a drawing on screen from AutoCAD 2012 can be carried out from either **Model Space** or from **Paper Space**.

## FIRST EXAMPLE – PRINTING (FIG. 15.36)

This example is of a drawing which has been acted upon by the **Visual Styles/Realistic** shading mode.

1. With a drawing to be printed or plotted on screen *click* the **Plot** tool icon in the **Output/Plot** panel (Fig. 15.34).
2. The **Plot** dialog appears (Fig. 15.35). Set the **Printer/Plotter** to a printer or plotter currently attached to the computer and the **Paper Size** to a paper size to which the printer/plotter is set.

**Fig. 15.34** The **Plot** icon in the **Output/Plot** panel.

**Fig. 15.35** The **Plot** dialog.

3.  *Click* the **Preview** button of the dialog and if the preview is OK (Fig. 15.36), *right-click* and in the *right-click* menu which appears, *click* **Plot**. The drawing plots producing the necessary "hardcopy".

## SECOND EXAMPLE – MULTIPLE VIEW COPY (FIG. 15.37)

The 3D model to be printed is a **Realistic** view of a 3D model. To print a multiple view copy:

1.  Place the drawing in a **Four: Equal** viewport setting.
2.  Make a new layer **vports** of color cyan and make it the current layer.

**Fig. 15.36** First example – **Print Preview** – **printing a single copy**.

**3.** *Click* the **Layout** button in the status bar. At the command line:

```
Command: enter mv (MVIEW) right-click
MVIEW
Specify corner of viewport or [ON/OFF/Fit/Shadeplot/
 Lock/Object/Polygonal/Restore/LAyer/2/3/4] <Fit>:
 enter r Restore) right-click
Enter viewport configuration name or [?] <*Active>:
 right-click
Specify first corner or [Fit] <Fit>: right-click
Command:
```

The drawing appears in **Paper Space**. The views of the 3D model appear each within a cyan outline in each viewport.

**4.** Turn layer **vports** off. The cyan outlines of the viewports disappear.

**5.** *Click* the **Plot** tool icon in the **Output/Plot** toolbar. Make sure the correct **Printer/Plotter** and **Paper Size** settings are selected and *click* the **Preview** button of the dialog.

**6.** If the preview is satisfactory (Fig. 15.37), *right-click* and from the *right-click* menu *click* **Plot**. The drawing plots to produce the required four-viewport hardcopy.

## SAVING AND OPENING 3D MODEL DRAWINGS

3D model drawings are saved and/or opened in the same way as are 2D drawings. To save a drawing *click* **Save As...** in the **File** drop-down menu and save the drawing in the **Save Drawing As**

**Fig. 15.37** Second example – **multiple view copy**.

dialog by *entering* a drawing file name in the **File Name** field of the dialog before *clicking* the **Save** button. To open a drawing that has been saved *click* **Open.** Go in the **File** drop-down menu and then select the **Select File** dialog. From there select the file that you are searching for by its name on the file list.

There are differences between saving a 2D and a 3D drawings. A 3D model drawing is shaded by using a visual style from the **Home/View** panel; the shading will be saved along with the drawing.

## CAMERA

## EXAMPLE – CAMERA SHOT IN ROOM SCENE

This example is of a camera being used in a room in which several chairs, stools, and tables have been placed. Start by constructing one of the chairs.

### Constructing One of the Chairs

1. In a **Top** view construct a polyline from an ellipse (after setting **pedit** to **1**) trimmed in half, then offset, and formed into a single pline using **pedit**.
2. Construct a polyline from a similar ellipse trimmed in half and then formed into a single pline using **pedit**.
3. Extrude both plines to suitable heights to form the chair frame and its cushion seat.

**4.** In a **Right** view, construct plines for the holes through the chair, extrude them to a suitable height, and subtract them from the extrusion of the chair frame.

**5.** Add suitable materials and render the result (Fig. 15.38).

**Fig. 15.38** Stages in constructing a chair.

## Constructing One of the Stools

**1.** In the **Front** view and working to suitable sizes, construct a pline outline for one quarter of the stool.

**2.** Extrude the pline to a suitable height.

**3.** **Mirror** the extrusion, followed by forming a union of the two mirrored parts.

**4.** In the **Top** view, copy the union, rotate the copy through 90 degrees and move it into a position across the original, and form a union of the two.

**5.** Add a cylindrical cushion and render (Fig. 15.39).

Pline for
extrusion

Extrusion          After Mirror          Seat added and
                   and Union             rendered

**Fig. 15.39** Stages in constructing a stool.

## Constructing One of the Tables

1. In the **Top** view and working to suitable sizes, construct a cylinder for the table top.
2. Construct two cylinders for the table rail and subtract the smaller from the larger.
3. Construct an ellipse from which a leg can be extruded and copy the extrusion three times to form the four legs.
4. In the **Front** view, move the parts to their correct positions relative to each other.
5. Add suitable materials and render (Fig. 15.40).

**Fig. 15.40** A Conceptual shading of one of the tables.

## Constructing Walls, Doors, and Window

Working to suitable sizes, construct walls, floor, doors, and window using the **Box** tool (Fig. 15.41).

**Fig. 15.41** A Conceptual style view of the walls, floor, doors, and window.

## USING A CAMERA

### Inserting the Furniture

In the **Top** view:

1. Insert the chair, copy it three times, and move the copies to suitable positions.
2. Insert the stool, copy it three times, and move the copies to suitable positions.
3. Insert the table, copy it three times, and move the copies to suitable positions (Fig. 15.42).

**Fig. 15.42 Top** view of the furniture inserted, copied and placed in position.

## Adding Lights

1. Place a **59 W 8 ft fluorescent** light central to the room just below the top of the wall height.
2. Place a **Point** light in the bottom right-hand central corner of the room (Fig. 15.43).

**Fig. 15.43** Two lights placed in the room.

## Placing a Camera

**Fig. 15.44** Selecting **Create Camera** from the **View** drop-down menu.

1. Place the scene in the **Front** view.
2. Select **Create Camera** from the **Render/Camera** panel, or from the **View** drop-down menu (Fig. 15.44).

The command line shows:

```
Command: _camera
Current camera settings: Height=0 Lens Length=80mm
Specify camera location: pick a position
Specify target location: drag to end of the cone
  into position
Enter an option [?/Name/LOcation/Height/Target/LEns/
  Clipping/View/eXit]<eXit>: enter
le (LEns) right-click
Specify lens length in mm <80>: enter 55 right-click
Enter an option [?/Name/LOcation/Height/Target/LEns/
  Clipping/View/eXit]<eXit>: n
```

```
Enter name for new camera <Camera2>: right-click
 -accepts name (Camera1)
Enter an option [?/Name/LOcation/Height/Target/LEns/
 Clipping/View/eXit]<eXit>: right-click
Command:
```

And the camera will be seen in position (Fig. 15.45).

**Fig. 15.45** The camera in position.

3. At the command line *enter* view. The **View Manager** dialog appears (Fig. 15.46).

**Fig. 15.46** Selecting **Camera1** from the **View Manager**.

In the **Views** list *click* **Camera1**, followed by a *click* on the **Set Current** button and then the **OK** button. A view of the camera view fills the AutoCAD drawing area.

4. If not satisfied with the scene it can be amended in several ways from the **Camera/Swivel** command (**View** drop-down menu) and its *right-click* menu (Figs 15.47 and 15.48).

**Fig. 15.47** Selecting **Camera/Swivel** from the **View** drop-down menu.

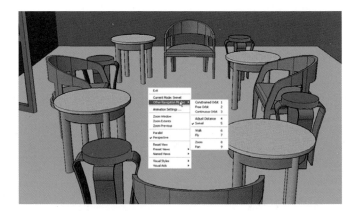

**Fig. 15.48** The camera view (**Conceptual**) after amendment and before render.

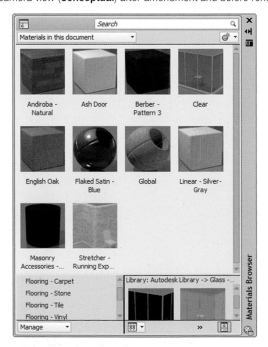

**Fig. 15.49** The materials in the scene in the **Materials** palette.

## OTHER FEATURES OF THIS SCENE

1. A fair number of materials were attached to objects as shown in the **Materials Browser** palette associated with the scene (Fig. 15.49).
2. Changing the lens to different lens lengths can make appreciable differences to the scene. One rendering of the same room scene taken with a lens of **55 mm** is shown in Fig. 15.50 and another with a **100 mm** lens is shown in Fig. 15.51.

**Fig. 15.50** The rendering of the scene taken with a **55 mm** lens.

**Fig. 15.51** The rendering of a scene taken with a **100 mm** lens camera.

## REVISION NOTES

1. 3D models can be constructed in any of the workspaces – 2D Design & Annotation, 3D Basics, or 3D Modeling. In Part B of this book, 3D models are constructed in either in the 3D Basics or the 3D Modeling workspace.

2. 3D model drawings can be constructed in either a Parallel projection or in a Perspective projection layout.

3. Materials and light palettes can be selected from the Render panels.

4. Materials can be modified from the Materials Editor palette.

5. In this book lighting of a scene with 3D models is mostly done by placing two distant lights in front of and above the models. One is positioned to the left and the other is placed to the right at a point above the center of the scene. The exception is the lighting of the camera scenes on pages 310–313.

6. There are many other methods of lighting a scene, in particular using default lighting or sun lighting.

7. Several Render preset methods of rendering are available, from Draft to Presentation.

8. The use of the Orbit tools allows a 3D model to be presented in any position.

9. Plotting or printing of either Model or Layout windows is possible.

10. Hardcopy can be from a single viewport or from multiple viewports. When printing or plotting 3D model drawings, Visual Style layouts print as they appear on screen.

## EXERCISES

Methods of constructing answers to the following exercises can be found in the free website:
**http://www.routledge.com/cw/yarwood**

1.  A rendering of an assembled lathe tool holder is shown in Fig. 15.52. The rendering includes different materials for each part of the assembly. Working to the dimensions given in the parts orthographic drawing (Fig. 15.53), construct a 3D model drawing of the assembled lathe tool holder on several layers of different colors, add lighting and materials and render the model in an isometric view. Shade with **3D Visual Styles/Hidden** and print or plot a **ViewCube/ Isometric** view of the model drawing.

**Fig. 15.52** Exercise 1.

**Fig. 15.53** Exercise 1 – parts drawings.

2.  Fig. 15.54 shows a rendering of a drip tray. Working to the sizes given in Fig. 15.55, construct a 3D model drawing of the tray. Add lighting and a suitable material; then place the model in an isometric view and render.

**Fig. 15.54** Exercise 2.

**Fig. 15.55** Exercise 2.

**3.** A three-view drawing of a hanging spindle bearing in third angle orthographic projection is shown in Fig. 15.56. Working to the dimensions in the drawing construct a 3D model drawing of the bearing. Add lighting and a material and render the model.

**Fig. 15.56** Exercise 3.

CHAPTER **16**

# BUILDING DRAWING

## AIMS OF THIS CHAPTER

The aims of this chapter are:

1. To show that AutoCAD 2012 is a suitable computer aided design software package for the construction of a building drawing.
2. To show that AutoCAD 2012 is a suitable CAD program for the construction of 3D models of a building.

## BUILDING DRAWINGS

There are a number of different types of drawings related to the construction of any form of building. In this chapter a fairly typical example of a set of building drawings is shown. These are seven drawings related to the construction of an extension to an existing two-story house (44 Ridgeway Road).

1. A site plan of the original two-story house, drawn to a scale of **1:200** (Fig. 16.1).
2. A site layout plan of the original house, drawn to a scale of **1:100** (Fig. 16.2).
3. Floor layouts of the original house, drawn to a scale of **1:50** (Fig. 16.3).
4. Views of all four sides of the original house, drawn to a scale of **1:50** (Fig. 16.4).
5. Floor layouts including the proposed extension, drawn to a scale of **1:50** (Fig. 16.5).

Introduction to AutoCAD 2012. ISBN: 978-0-08-096947-3

**Fig. 16.1** A site plan.

**Fig. 16.2** A site layout plan.

**Fig. 16.3** Floor layouts drawing of the original house.

**Fig. 16.4** Views of the original house.

6. Views of all four sides of the house including the proposed extension, drawn to a scale of **1:50** (Fig. 16.6).

7. A sectional view through the proposed extension, drawn to a scale of **1:50** (Fig. 16.7).

**Fig. 16.5** Floor layouts drawing of the proposed extension.

**Fig. 16.6** Views including the proposed extension.

NOTES →

1. Other types of drawings will be constructed such as drawings showing the details of parts such as doors, windows, floor structures etc. These are often shown in sectional views.

2. Although the seven drawings related to the proposed extension of the house at 44 Ridgeway Road are shown here as having been constructed on either A3 or A4 layouts, it is common practice to include several types of building drawings on larger sheets such as A1 sheets of a size 820 mm by 594 mm.

**Fig. 16.7** A section through the proposed extension.

## FLOOR LAYOUTS

When constructing floor layout drawings it is advisable to build up a library of block drawings of symbols representing features such as doors and windows. These can then be inserted into layouts from the **DesignCenter**. A suggested small library of such block symbols is shown in Fig. 16.8.

**Fig. 16.8** A small library of building symbols.

Details of shapes and dimensions for the first two examples are taken from the drawings of the building and its extension at 44 Ridgeway Road given in the Figs 16.2 to 16.7.

## 3D MODELS OF BUILDINGS

Details of this first example are taken from Figs 16.2 to 16.4 on pages 318 and 319.

The following steps describe the construction of a 3D model of 44 Ridgeway Road prior to the extension being added.

### FIRST EXAMPLE – 44 RIDGEWAY ROAD – ORIGINAL BUILDING (FIG. 16.19)

1. In the **Layer Properties Manager** palette – **Doors** (color **red**), **Roof** (color **green**), **Walls** (color **blue**), **Windows** (color **Magenta**) (Fig.16.9).

**Fig. 16.9** First example – the layers on which the model is to be constructed.

**Fig. 16.10** Set screen to the **ViewCube/Front** view.

2. Set the screen to the **ViewCube/Front** view (Fig. 16.10)
3. Set the layer **Walls** current and, working to a scale of **1:50** construct outlines of the walls. Construct outlines of the bay, windows, and doors inside the wall outlines.
4. **Extrude** the wall, bay, window, and door outlines to a height of **1**.
5. **Subtract** the bay, window, and door outlines from the wall outlines. The result is shown in Fig. 16.11.
6. Make the layer **Windows** current and construct outlines of three of the windows which are of different sizes. Extrude the copings and cills to a height of **1.5** and the other parts to a height of **1**. Form a union of the main outline, the coping, and the cill. The window pane extrusions will have to be subtracted from the union. Fig. 16.12 shows the 3D models of the three windows in a **ViewCube/Isometric** view.
7. Move and copy the windows to their correct positions in the walls.
8. Make the layer **Doors** current and construct outlines of the doors and extrude to a height of **1**.

**Fig. 16.11** First example – the walls.

**Fig. 16.12** First example – extrusions of the three sizes of windows.

**9.** Make layer **Chimney** current and construct a 3D model of the chimney (Fig. 16.13).

**10.** Make the layer **Roofs** current and construct outlines of the roofs (main building and garage). See Fig. 16.14.

**Fig. 16.13** First example – Realistic view of a 3D model of the chimney.

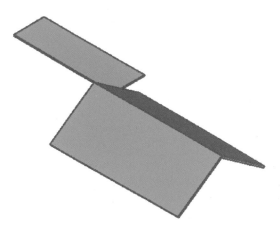

**Fig. 16.14** First example – **Realistic** view of the roofs.

**11.** On the layer **Bay** construct the bay and its windows.

## Assembling the Walls

**Fig. 16.15** Set screen to ViewCube/**Top** view.

1. Place the screen in the **ViewCube/Top** view (Fig. 16.15).
2. Make the layer **Walls** current and turn off all layers other than **Windows**.
3. Placing a window around each wall in turn, move and/or rotate and move the walls until they are in their correct position relative to each other.
4. Place in the **ViewCube/Isometric** view and, using the **Move** tool, move the walls into their correct positions relative to each other. Fig. 16.16 shows the walls in position in a **ViewCube/Top** view.

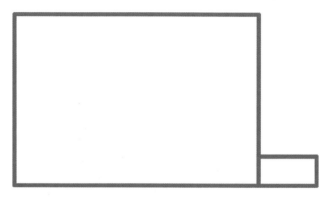

**Fig. 16.16** First example – the four walls in their correct positions relative to each other in a **ViewCube/Top** view.

5. Move the roof into position relative to the walls and move the chimney into position on the roof. Fig. 16.17 shows the resulting 3D model in a **ViewCube/Isometric** view (Fig. 16.18).

**Fig. 16.18** Set screen to a **ViewCube/Isometric** view.

**Fig. 16.17** First example – a **Realistic** view of the assembled walls, windows bay, roof, and chimney.

## The Garage

On layer **Walls** construct the walls and on layer **Windows** construct the windows. Fig. 16.19 shows a **Realistic** visual style view of the 3D model as constructed so far.

**Fig. 16.19** First example – **Realistic** view of the original house and garage.

## SECOND EXAMPLE – EXTENSION TO 44 RIDGEWAY ROAD (FIG. 16.20)

Working to a scale of **1:50** and taking dimensions from the drawings in Figs 16.5 and 16.6 and in a manner similar to the method of constructing the 3D model of the original building, add the extension to the original building. Fig. 16.20 shows a **Realistic** visual style view of the resulting 3D model. In this 3D model floors have been added – a ground and a first story floor constructed on a new layer **Floors** of color yellow. Note the changes in the bay and front door.

## THIRD EXAMPLE – SMALL BUILDING IN FIELDS (FIG. 16.23)

Working to a scale of **1:50** from the dimensions given in Fig. 16.21, construct a 3D model of the hut using the steps given in the following text.

**Fig. 16.20** Second example – a **Realistic** view of the building with its extension.

**Fig. 16.21** Third example – front and end views of the hut.

In the **Layer Properties Manager** dialog make the new levels as follows:

**Walls** – color **Blue**
**Road** – color **Red**
**Roof** – color **Red**
**Windows** – **Magenta**
**Fence** – color **8**
**Field** – color **Green**

Following the methods used in the construction of the house in the first example, construct the walls, roof, windows, and a door of the small building in one of the fields. Fig. 16.22 shows a **Realistic** visual style view of a 3D model of the hut.

**Fig. 16.22** Third example – a **Realistic** view of a 3D model of the hut.

## Constructing the Fence, Fields, and Road

1. Place the screen in a **Four: Equal** viewports setting.
2. Make the **Garden** layer current and, in the **Top** viewport, construct an outline of the boundaries to the fields and to the building. Extrude the outline to a height of **0.5**.
3. Make the **Road** layer current and in the **Top** viewport, construct an outline of the road and extrude the outline to a height of **0.5**.
4. In the **Front** view, construct a single plank and a post of a fence and copy them a sufficient number of times to surround the four fields leaving gaps for the gates. With the **Union** tool form a union of all the posts and planks. Fig. 16.23 shows a part of the resulting fence in a **Realistic** visual style view in the **Isometric** viewport. With the **Union** tool form a union of all the planks and posts in the entire fence.

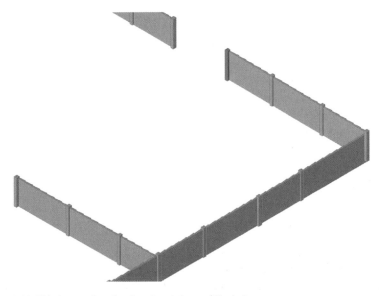

**Fig. 16.23** Third example – front and end views of the hut.

5. While still in the layer **Fence**, construct gates to the fields.
6. Make the **Road** layer current and construct an outline of the road. Extrude to a height of **0.5**.

> **NOTE** →
>
> When constructing each of these features it is advisable to turn off those layers on which other features have been constructed.
>
> Fig. 16.24 shows a **Conceptual** view of the hut in the fields with the road, fence, and gates.
>
>
>
> **Fig. 16.24**  Third example – the completed 3D model, **Conceptual** view.

## Completing the Second Example

Working in a manner similar to the method used when constructing the roads, garden, and fences for the third example, add the paths, garden area and fences and gates to the building 44 Ridgeway Road with its extension. Fig. 16.25 shows a **Conceptual** visual style view of the resulting 3D model.

## MATERIAL ATTACHMENTS AND RENDERING

### SECOND EXAMPLE – EXTENSION TO 44 RIDGEWAY ROAD

The following materials were attached to the various parts of the 3D model (Fig. 16.26).

To attach the materials, all layers except the layer on which the objects to which the attachment of a particular material is being

**Fig. 16.25** Second example – the completed 3D model.

**Fig. 16.26** Second example – 3D model after attaching materials and rendering.

made are turned off, allowing the material in question to be attached only to the elements to which each material is to be attached.

**Default:** color 7.
**Doors:** Wood Hickory
**Fences:** Wood – Spruce
**Floors:** Wood – Hickory
**Garden:** Green
**Gates:** Wood – White
**Roofs:** Brick – Herringbone
**Windows:** Wood – White

The 3D model was then rendered with **Output Size** set to **1024x768** and **Render Preset** set to **Presentation**, with **Sun Status** turned on. The resulting rendering is shown in Fig. 16.26.

## THIRD EXAMPLE – SMALL BUILDING IN FIELDS

Fig. 16.27 shows the third example after attaching materials and rendering.

**Fig. 16.27** Third example – the completed 3D model.

## REVISION NOTES ↻

1. There are a number of different types of building drawings – site plans, site layout plans, floor layouts, views, sectional views, detail drawings. AutoCAD 2012 is a suitable CAD program to use when constructing building drawings.

2. AutoCAD 2012 is a suitable CAD program for the construction of 3D models of buildings.

EXCERCISES (vertical, right margin)

## EXERCISES

Methods of constructing answers to the following exercises can be found in the free website:

**http://www.routledge.com/cw/yarwood**

**1.** Fig. 16.28 is a site plan drawn to a scale of 1:200 showing a bungalow to be built in the garden of an existing bungalow. Construct the library of symbols shown in Fig. 16.8 on page 321 and by inserting the symbols from the DesignCenter construct a scale 1:50 drawing of the floor layout plan of the proposed bungalow.

**Fig. 16.28** Exercise 1.

**2.** Fig. 16.29 shows a site plan of a two-story house in a building plot. Design and construct, to a scale 1:50, a suggested pair of floor layouts for the two floors of the proposed house.

**Fig. 16.29** Exercise 2.

**3.** Fig. 16.30 shows a scale 1:100 site plan for the proposed bungalow 4 Caretaker Road. Construct the floor layout for the proposed house shown in the drawing in Fig. 16.26.

**Fig. 16.30** Exercise 3 – site plan.

**4.** Fig. 16.31 shows a building plan of a house in a site plan (Fig. 16.28). Construct a 3D model view of the house making an assumption as to the roofing and the heights connected with your model.

**Fig. 16.31** Exercise 3 – a building.

**5.** Fig. 16.32 shows a three view, dimensioned orthographic projection of a house. Fig.16.33 is a rendering of a 3D model of the house. Construct the 3D model to a scale of 1:50, making estimates of dimensions not given in Fig. 16.32 and render using suitable materials.

**Fig. 16.32** Exercise 5 – orthographic views.

**Fig. 16.33** Exercise 5 – the rendered model.

**6.** Fig. 16.34 shows a two view orthographic projection of a small garage. Fig. 16.35 shows a rendering of a 3D model of the garage. Construct the 3D model of the garage working to a suitable scale.

**Fig. 16.34** Exercise 6 – orthographic views.

**Fig. 16.35** Exercise 6.

**7.** Fig. 16.36 shows a two-view orthographic projection of a garden seat and Fig. 16.37 a 3D solid model drawing of the garden seat displayed in the **Visual Style Shaded**. Working to a suitable scale, construct the 3D solid model drawing, working to the dimensions given in Fig. 16.36.

**Fig. 16.36** Exercise 7 – orthographic views.

**Fig. 16.37** Exercise 7.

**8.** Three orthographic projections of garden tables are shown in Figs 16.38 to Fig. 16.40. Fig. 16.41 shows 3D models of the tables.

**Fig. 16.38** Exercise 8 – Garden table No. 1.

**Fig. 16.39** Exercise 8 – Garden table No. 2.

Garden table - No. 3

**Fig. 16.40** Exercise 8 – Garden table No. 3.

Garden table No. 1

Garden table No. 2

Garden table No. 3

**Fig. 16.41** Exercise 8 – 3D models of the three Garden tables.

Choose the table you prefer from the three. Give reasons for you choice. Then construct a 3D model of the table of your choice. Render your drawing using a Paint material of an appropriate color.

# THREE-DIMENSIONAL SPACE

## AIM OF THIS CHAPTER

The aim of this chapter is to show in examples the methods of manipulating 3D models in 3D space using the **UCS** tools from the **View/Coordinates** panel or from the command line.

## 3D SPACE

So far in this book, when constructing 3D model drawings, they have been constructed on the AutoCAD 2012 coordinate system which is based upon three planes:

The **XY Plane** – the screen of the computer.

The **XZ Plane** at right angles to the **XY Plane** and as if coming towards the operator of the computer.

A third plane (**YZ**) lying at right angles to the other two planes (Fig. 17.1).

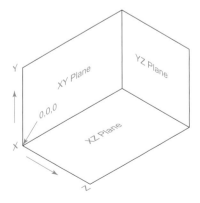

**Fig. 17.1** The 3D space planes.

**Introduction to AutoCAD 2012. ISBN: 978-0-08-096947-3**

In earlier chapters the **3D Navigate** drop-down menu and the **ViewCube** have been described to enable 3D objects which have been constructed on these three planes to be viewed from different viewing positions. Another method of placing the model in 3D space using the **Orbit** tool has also been described.

## THE USER COORDINATE SYSTEM (UCS)

The **XY** plane is the basic **UCS** plane, which in terms of the ucs is known as the \*WORLD\* plane.

The **UCS** allows the operator to place the AutoCAD coordinate system in any position in 3D space using a variety of **UCS** tools (commands). Features of the **UCS** can be called either by *entering* **ucs** at the command line or by the selection of tools from the **View/Coordinates** panel (Fig. 17.2). Note that a *click* on **World** in the panel brings a drop-down menu from which other views can be selected (Fig. 17.3).

**Fig. 17.2** The **View/Coordinates** panel.

**Fig. 17.3** The drop-down menu from **World** in the **View/Coordinates** panel.

If **ucs** is *entered* at the command line, it shows:

```
Command: enter ucs right-click
Current ucs name: *WORLD*
Specify origin of UCS or [Face/NAmed/OBject/
  Previous/View/World/X/Y/Z/ZAxis] <World>:
```

And from these prompts selection can be made.

## THE VARIABLE UCSFOLLOW

UCS planes can be set from using the methods shown in Figs 17.2 and 17.3 or by *entering* **ucs** at the command line. No matter which method is used, the variable **UCSFOLLOW** must first be set on as follows:

```
Command: enter ucsfollow right-click
Enter new value for UCSFOLLOW <0>: enter 1 right-click
Command:
```

## THE UCS ICON

The **UCS** icon indicates the directions in which the three coordinate axes **X, Y,** and **Z** lie in the AutoCAD drawing. When working in 2D, only the **X** and **Y** axes are showing, but when the drawing area is in a 3D view all three coordinate arrows are showing, except when the model is in the **XY** plane. The icon can be turned off as follows:

```
Command: enter ucsicon right-click
Enter an option [ON/OFF/All/Noorigin/ORigin/
  Properties] <ON>:
```

To turn the icon off, *enter* **off** in response to the prompt line and the icon disappears from the screen.

The appearance of the icon can be changed by *entering* **p** (Properties) in response to the prompt line. The **UCS Icon** dialog appears in which changes can be made to the shape, line width, and color of the icon if wished.

### Types of UCS Icon

The shape of the icon can be varied partly when changes are made in the **UCS Icon** dialog but also according to whether the AutoCAD drawing area is in 2D, 3D, or PaperSpace (Fig. 17.4).

2D icon    3D icon    3D icon    PaperSpace
                                     icon

**Fig. 17.4** Types of **UCS** icon.

## EXAMPLES OF CHANGING PLANES USING THE UCS

### FIRST EXAMPLE – CHANGING UCS PLANES (FIG. 17.6)

1. Set **UCSFOLLOW** to **1** (ON).
2. Make a new layer color **Red** and make the layer current. Place the screen in **ViewCube/Front** and **Zoom** to **1**.
3. Construct the pline outline in Fig. 17.5 and extrude to **120** high.

**Fig. 17.5** First example – **Changing UCS planes** – pline for extrusion.

4. Place in **ViewCube/Isometric** view and **Zoom** to **1**.
5. With the **Fillet** tool, fillet corners to a radius of **20**.
6. At the command line:

```
Command: enter ucs right-click
Current ucs name: *WORLD*
Specify origin of UCS or [Face/NAmed/OBject/
  Previous/View/World/X/Y/Z/ZAxis] <World>: enter f
  (Face) right-click
Select face of solid object: pick the sloping face -
  its outline highlights
Enter an option [Next/Xflip/Yflip] <accept>: right-
  click
Regenerating model
Command:
```

And the 3D model changes its plane so that the sloping face is now on the new UCS plane. **Zoom** to **1**.
7. On this new UCS, construct four cylinders of radius 7.5 and height -15 (note the minus) and subtract them from the face.
8. *Enter* **ucs** at the command line again and *right-click* to place the model in the *WORLD* UCS.

9. Place four cylinders of the same radius and height into position in the base of the model and subtract them from the model.
10. Place the 3D model in a **ViewCube/Isometric** view and set in the **Home/View/Conceptual** visual style (Fig. 17.6).

## SECOND EXAMPLE – UCS (FIG. 17.11)

The 3D model for this example is a steam venting valve – a two-view third angle projection of the valve is shown in Fig. 17.7.

**Fig. 17.6** First example – **Changing UCS planes**.

**Fig. 17.7** Second example – **UCS** – The orthographic projection of a steam venting valve.

1. Make sure that **UCSFOLLOW** is set to **1**.
2. Place in the **UCS \*WORLD\*** view. Construct the **120** square plate at the base of the central portion of the valve. Construct five cylinders for the holes in the plate. Subtract the five cylinders from the base plate.
3. Construct the central part of the valve – a filleted **80** square extrusion with a central hole.
4. At the command line:

```
Command: enter ucs right-click
Current ucs name: *WORLD*
Specify origin of UCS or
[Face/NAmed/OBject/Previous/View/World/X/Y/Z/ZAxis]
  <World>: enter x right-click
```

**Fig. 17.8** Second example – **USC** – step **11** + rendering.

```
Specify rotation angle about X axis <90>: right-click
Command:
```

And the model assumes a **Front** view.

5. With the **Move** tool, move the central portion vertically up by **10**.
6. With the **Copy** tool, copy the base up to the top of the central portion.
7. With the **Union** tool form a single 3D model of the three parts.
8. Make the layer **Construction** current.
9. Place the model in the **UCS \*WORLD\*** view. Construct the separate top part of the valve – a plate forming a union with a hexagonal plate and with holes matching those of the other parts.
10. Place the drawing in the **UCS X** view. Move the parts of the top into their correct positions relative to each other. With **Union** and **Subtract** complete the part. This will be made easier if the layer 0 is turned off.

**Fig. 17.9** Second example – USC – steps **12** and **13** + rendering.

11. Turn layer 0 back on and move the top into its correct position relative to the main part of the valve. Then with the **Mirror** tool, mirror the top to produce the bottom of the assembly.
12. While in the **UCS X** view construct the three parts of a 3D model of the extrusion to the main body.
13. In the **UCS \*WORLD\*** view, move the parts into their correct position relative to each other. **Union** the two filleted rectangular extrusions and the main body. **Subtract** the cylinder from the whole.

**Fig. 17.10** Second example – UCS – pline for the bolt.

14. In the **UCS X** view, construct one of the bolts as shown in Fig. 17.10, forming a solid of revolution from a pline. Then construct a head to the bolt and with **Union** add it to the screw.
15. With the **Copy** tool, copy the bolt 7 times to give 8 bolts. With **Move**, and working in the **UCS \*WORLD\*** and **X** views, move the bolts into their correct positions relative to the 3D model.
16. Add suitable lighting and attach materials to all parts of the assembly and render the model.
17. Place the model in the **ViewCube/Isometric** view.
18. Save the model to a suitable file name.
19. Finally move all the parts away from each other to form an exploded view of the assembly (Fig. 17.11).

**Fig. 17.11** Second example – **UCS**.

## THIRD EXAMPLE – UCS (FIG. 17.15)

1. Set **UCSFOLLOW** to 1.
2. Place the drawing area in the **UCS X** view.

3. Construct the outline in Fig. 17.12 and extrude to a height of **120**.
4. *Click* the **3 Point** tool icon in the **View/Coordinates** panel (Fig. 17.13):

**Fig. 17.12** Third example – **UCS** – outline for 3D model.

**Fig. 17.13** The **UCS, 3 Point** icon in the **View/Coordinates** panel.

```
Command: _ucs
Current ucs name: *WORLD*
Specify origin of UCS or [Face/NAmed/OBject/
 Previous/View/World/X/Y/Z/ZAxis] <World>: _3
Specify new origin point <0,0,0>: pick point (Fig.
 17.14)
Specify point on positive portion of X-axis: pick
 point (Fig. 17.14)
Specify point on positive-Y portion of the UCS XY
 plane <-142,200,0>: enter .xy right-click
of pick new origin point (Fig. 17.14) (need Z):
 enter 1 right-click
Regenerating model
Command:
```

Fig. 17.14 shows the **UCS** points; the model regenerates in this new 3point plane.

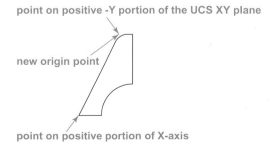

point on positive -Y portion of the UCS XY plane

new origin point

point on positive portion of X-axis

**Fig. 17.14** Third example – **UCS** – the three UCS points.

**Fig. 17.15** Third example – **UCS**.

5. On the face of the model construct a rectangle 80 x 50 central to the face of the front of the model, fillet its corners to a radius of 10 and extrude to a height of **10**.
6. Place the model in the **ViewCube/Isometric** view and fillet the back edges of the second extrusion to a radius of **10**.
7. Subtract the second extrusion from the first.
8. Add lights and a suitable material and render the model (Fig. 17.15).

## FOURTH EXAMPLE – UCS (FIG. 17.17)

1. With the last example still on screen, place the model in the **UCS *WORLD*** view.
2. Call the **Rotate** tool from the **Home/Modify** panel and rotate the model through 225 degrees.
3. *Click* the **X** tool icon in the **View/Coordinates** panel (Fig. 17.16):

**Fig. 17.16** The **UCS X** tool in the **View/Coordinates** panel.

```
Command: _ucs
Current ucs name: *WORLD*
Specify origin of UCS or [Face/NAmed/OBject/
  Previous/View/World/X/Y/Z/ZAxis] <World>: _x
Specify rotation angle about X axis <90>: right-click
Regenerating model
Command:
```

**Fig. 17.17** Fourth example.

4. Render the model in its new **UCS** plane (Fig. 17.17).

## SAVING UCS VIEWS

If a number of different **UCS** planes are used in connection with the construction of a 3D model, each view obtained can be saved to a different name and recalled when required. To save a UCS plane view in which a 3D model drawing is being constructed *enter* **ucs** at the command line:

```
Current ucs name: *NO NAME*
Specify origin of UCS or [Face/NAmed/OBject/
  Previous/View/World/X/Y/Z/ZAxis] <World>: enter s
  right-click
Enter name to save current UCS or [?]: enter New
  View right-click
Regenerating model
Command:
```

*Click* the **UCS Settings** arrow in the **View/Coordinates** panel and the **UCS** dialog appears. *Click* the **Named UCSs** tab of the dialog and the names of views saved in the drawing appear (Fig. 17.18).

**Fig. 17.18** The **UCS** dialog.

## CONSTRUCTING 2D OBJECTS IN 3D SPACE

In previous chapters, there have been examples of 2D objects constructed with the **Polyline, Line, Circle,** and other 2D tools to form the outlines for extrusions and solids of revolution. These outlines have been drawn on planes in the **ViewCube** settings.

## FIRST EXAMPLE – 2D OUTLINES IN 3D SPACE (FIG. 17.21)

1. Construct a **3point UCS** to the following points:
```
Origin point: 80,90
X-axis point: 290,150
Positive-Y point: .xy of 80,90
(need Z): enter 1
```
2. On this **3point UCS** construct a 2D drawing of the plate to the dimensions given in Fig. 17.19, using the **Polyline, Ellipse**, and **Circle** tools.

Holes Ø20

All chamfers are 10x10

**Fig. 17.19** First example – **2D outlines in 3D space**

**Fig. 17.20** First example – **2D outlines in 3D space**. The outline in the **Isometric** view.

3. Save the **UCS** plane in the **UCS** dialog to the name **3point**.
4. Place the drawing area in the **ViewCube/Isometric** view (Fig. 17.20).
5. Make the layer **Red** current
6. With the **Region** tool form regions of the 6 parts of the drawing and, with the **Subtract** tool, subtract the circles and ellipse from the main outline.
7. Place in the **View/Visual Style/Realistic** visual style. Extrude the region to a height of **10** (Fig. 17.21).

## SECOND EXAMPLE – 2D OUTLINES IN 3D SPACE (FIG. 17.25)

1. Place the drawing area in the **ViewCube/Front** view, **Zoom** to **1** and construct the outline in Fig. 17.22.
2. Extrude the outline to **150** high.
3. Place in the **ViewCube/Isometric** view and **Zoom** to **1**.

**Fig. 17.21** First example – **2D outlines in 3D space**.

**Fig. 17.22** Second example – **2D outlines in 3D space**. Outline to be extruded.

4. *Click* the **Face** tool icon in the **View/Coordinates** panel (Fig. 17.23) and place the 3D model in the ucs plane shown in Fig. 17.24, selecting the sloping face of the extrusion for the plane and again **Zoom** to 1.

**Fig. 17.23** The **Face** icon from the **View/Coordinates** panel.

5. With the **Circle** tool draw five circles as shown in Fig. 17.24.

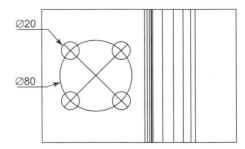

**Fig. 17.24** Second example – **2D outlines in 3D space**.

**Fig. 17.25** Second example – 2D outlines in 3D space.

6. Form a region from the five circles and with **Union** form a union of the regions.

7. Extrude the region to a height of −**60** (Note the minus) – higher than the width of the sloping part of the 3D model.

8. Place the model in the **ViewCube/Isometric** view and subtract the extruded region from the model.

9. With the **Fillet** tool, fillet the upper corners of the slope of the main extrusion to a radius of **30**.

10. Place the model into another **UCS FACE** plane and construct a filleted pline of sides **80** and **50** and filleted to a radius of **20**. Extrude to a height of −**60** and subtract the extrusion from the 3D model.

11. Place in the **ViewCube/Isometric** view, add lighting and a material.

## THE SURFACES TOOLS

The construction of 3D surfaces from lines, arc, and plines has been dealt with – see pages 224 to 276. In this chapter examples of 3D surfaces constructed with the tools **Edgesurf**, **Rulesurf**, and **Tabsurf** will be described. The tools can be called from the **Mesh/Primitives** panel. Fig. 17.26 shows the **Tabulated Surface** tool icon in the panel. The two icons to the right of that shown are the **Ruled Surface** and

**Fig. 17.26** The **Tabulated Surface** tool icon in the **Mesh/Primitives** panel.

the **Edge Surface** tools. In this chapter these three surface tools will be called by *entering* their tool names at the command line.

## Surface Meshes

Surface meshes are controlled by the set variables **Surftab1** and **Surftab2**. These variables are set as follows:

At the command line:

```
Command: enter surftab1 right-click
Enter new value for SURFTAB1 <6>: enter 24 right-click
Command:
```

## THE EDGESURF TOOL – FIG. 17.29

1. Make a new layer colour **magenta**. Make that layer current.
2. Place the drawing area in the **View Cube/Right** view. **Zoom** to **All**.
3. Construct the polyline to the sizes and shape as shown in Fig. 17.27.
4. Place the drawing area in the **View Cube/Top** view. **Zoom** to **All**.

**Fig. 17.27** Example – **Edgesurf –** pline outline.

5. Copy the pline to the right by **250**.
6. Place the drawing in the **ViewCube/Isometric** view. **Zoom** to **All**.
7. With the **Line** tool, draw lines between the ends of the two plines using the **endpoint** osnap (Fig. 17.28). Note that if polylines are drawn they will not be accurate at this stage.
8. Set **SURFTAB1** to **32** and **SURFTAB2** to **64**.
9. At the command line:

**Fig. 17.28** Example – **Edgesurf** – adding lines joining the plines.

```
Command: enter edgesurf right-click
Current wire frame density: SURFTAB1=32 SURFTAB2=64
Select object 1 for surface edge: pick one of the
  lines (or plines)
Select object 2 for surface edge: pick the next
  adjacent line (or pline)
Select object 3 for surface edge: pick the next
  adjacent line (or pline)
Select object 4 for surface edge: pick the last line
  (or pline)
Command:
```

The result is shown in Fig. 17.29.

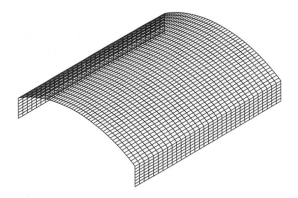

**Fig. 17.29** Example – **Edgesurf.**

**Fig. 17.30 Rulesurf** – the outline.

**Fig. 17.31** Example – **Rulesurf**.

## THE RULESURF TOOL – FIG. 17.31

1. Make a new layer color **blue** and make the layer current.
2. In the **ViewCube/Front** view construct the pline as shown in Fig. 17.30.
3. In the **3D Navigate/Top, Zoom** to 1 and copy the pline to a vertical distance of **120**.
4. Place in the **3D Navigate/Southwest Isometric** view and **Zoom** to 1
5. Set **SURFTAB1** to **32**.
6. At the command line:

```
Command: enter rulesurf right-click
Current wire frame density: SURFTAB1=32
Select first defining curve: pick one of the plines
Select second defining curve: pick the other pline
Command:
```

The result is given in Fig. 17.31.

## THE TABSURF TOOL – FIG. 17.32

1. Make a new layer of color **red** and make the layer current.
2. Set **Surftab1** to **2**.
3. In the **ViewCube/Top** view construct a hexagon of edge length 35.
4. In the **ViewCube/Front** view and in the center of the hexagon construct a pline of height **100**.

**5.** Place the drawing in the **ViewCube/Isometric** view.

**6.** At the command line:

```
Command: enter tabsurf right-click
Current wire frame density: SURFTAB1=2
Select objects for path curve: pick the hexagon
Select object for direction vector: pick the pline
Command:
```

See Fig. 17.32.

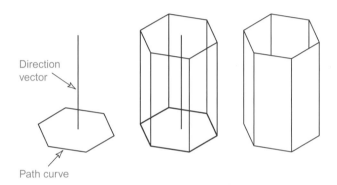

Direction vector

Path curve

**Fig. 17.32** Example – **Tabsurf**.

## REVISION NOTES

1. The UCS (User Coordinate System) tools can be called from the View/Coordinates panel or by entering ucs at the command line.
2. The variable UCSFOLLOW must first be set on (to 1) before operations of the UCS can be brought into action.
3. There are three types of UCS icon – 2D, 3D, and Paper Space.
4. The position of the plane in 3D space on which a drawing is being constructed can be varied using tools from the View/Coordinates panel.
5. The planes on which drawings are constructed on different planes in 3D space can be saved in the UCS dialog.
6. The tools Edgesurf, Rulesurf, and Tabsurf can be used to construct surfaces in addition to surfaces which can be constructed from plines and lines using the Extrude tool.

## EXERCISES

Methods of constructing answers to the following exercises can be found in the free website:
**http://www.routledge.com/cw/yarwood**

1.  Fig. 17.33 shows a rendering of a two-view projection of an angle bracket in which two pins are placed in holes in each of the arms of the bracket. Fig. 17.34 shows a two-view projection of the bracket. Construct a 3D model of the bracket and its pins. Add lighting to the scene, and materials to the parts of the model, and render.

2.  The two-view projection (Fig. 17.35) shows a stand consisting of two hexagonal prisms. Circular holes have been cut right through each face of the smaller hexagonal prism and rectangular holes with rounded ends have been cut right through the faces of the larger. Construct a 3D model of the stand. When completed add suitable lighting to the scene. Then add a material to the model and render (Fig. 17.36).

**Fig. 17.33** Exercise 1 – a rendering.

**Fig. 17.34** Exercise 1 – details of shape and sizes.

**Fig. 17.35** Exercise 2 – details of shapes and sizes.

**Fig. 17.36** Exercise 2 – a rendering.

3. The two-view projection in Fig. 17.37 shows a ducting pipe. Construct a 3D model drawing of the pipe. Place in an **SW Isometric** view, add lighting to the scene and a material to the model, and render.

**Fig. 17.37** Exercise 3 – details of shape and sizes.

4. A point marking device is shown in two two-view projections in Fig. 17.38. The device is composed of three parts – a base, an arm, and a pin. Construct a 3D model of the assembled device and add appropriate materials to each part. Then add lighting to the scene and render in an **SW Isometric** view (Fig. 17.39).

**Fig. 17.38** Exercise 4 – details of shapes and sizes.

**Fig. 17.39** Exercise 4 – a rendering.

**5.** A rendering of a 3D model drawing of the connecting device shown in the orthographic projection in Fig. 17.40 is given in Fig. 17.41. Construct the 3D model drawing of the device and add suitable lighting to the scene. Then place in the **ViewCube/Isometric** view, add a material to the model, and render.

Ø3.55"
Ø2.95"
R2.35"
R1.55"
Ø0.60
R0.59"

0.59"
R0.20"
0.20"
3.15"
0.30"

**Fig. 17.40** Exercise 5 – two-view drawing.

**Fig. 17.41** Exercise 5 – a rendering.

**6.** A fork connector and its rod are shown in a two-view projection in Fig. 17.42. Construct a 3D model drawing of the connector with its rod in position. Then add lighting to the scene, place in the **ViewCube/Isometric** viewing position, add materials to the model and render.

Hole Ø20    Ø30

R3
R35

45

140

100

R20    Ø20

15    40

**Fig. 17.42** Exercise 6.

**7.** An orthographic projection of the parts of a lathe steady are given in Fig. 17.43. From the dimensions shown in the drawing, construct an assembled 3D model of the lathe steady. When the 3D model has been completed, add suitable lighting and materials and render the model (Fig. 17.44).

Dimensions in millimeters    DO NOT SCALE THIS DRAWING

26
3

6
5
45
90°
30
BOSS Ø16
TAPPED M8
30
80°    R24
13
40
M8
116
80
5
2
123
1
30xØ10
12
Ø16    45    10    3
M10    ÿ20
10
10
HOLES Ø10
63    Ø20
40    10    4
A
98    13
M10    Ø20

Name: A. Student    Scale: 1:1    Date: 15/03/2008    Title: LATHE STEADY

**Fig. 17.43** Exercise 7 – details.

EXERCISES

**Fig. 17.44** Exercise 7 – a rendering.

8. Construct suitable polylines to sizes of your own discretion in order to form two surfaces to form the box shape shown in Fig. 17.45 with the aid of the **Rulesurf** tool. Add lighting and a material and render the surfaces so formed. Construct another three **Edgesurf** surfaces to form a lid for the box. Place the surface in a position above the box, add a material and render (Fig. 17.46)

**Fig. 17.45** Exercise 8 – the box.

**Fig. 17.46** Exercise 8 – the box and its lid.

9. Fig. 17.47 shows a polyline for each of the 4 objects from which the surface shown in Fig. 17.48 was obtained. Construct the surface and shade in **Shades of Gray**.

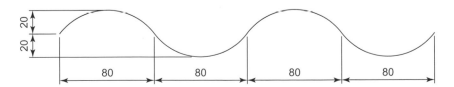

**Fig. 17.47** Exercise 9 – one of the polylines from which the surface was obtained.

**Fig. 17.48** Exercise 9.

**10.** The surface model for this exercise was constructed from 3 **Edgesurf** surfaces working to the suggested objects for the surface as shown in Fig. 17.51. The sizes of the outlines of the objects in each case are left to your discretion. Fig. 17.49 shows the completed surface model. Fig. 17.50 shows the three surfaces of the model separated from each other.

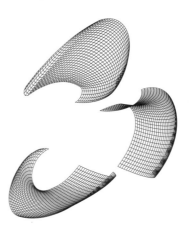

**Fig. 17.49** Exercise 10.

**Fig. 17.50** Exercise 10 – The three surfaces.

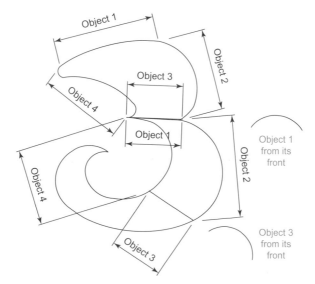

**Fig. 17.51** Exercise 10 – Outlines for the three surfaces.

11. Fig. 17.52 shows in a **View Block/Isometric** view a semicircle of radius **25** constructed in the **View Cube/Top** view on a layer of color **Magenta** with a semicircle of radius **75** constructed in the **View Block/Front** view with its left-hand end centered on the semicircle. Fig. 17.53 shows a surface constructed from the two semicircles in a **Visual Styles/Realistic** mode.

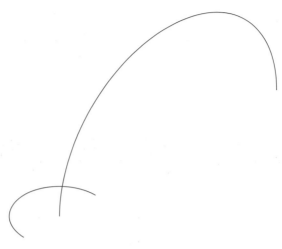

**Fig. 17.52** Exercise 11 – the circle and the semicircle.

**Fig. 17.53** Exercise 11.

CHAPTER **18**

# EDITING 3D SOLID MODELS

## AIMS OF THIS CHAPTER

The aims of this chapter are:
1. To introduce the use of tools from the **Solid Editing** panel.
2. To show examples of a variety of 3D solid models.

## THE SOLID EDITING TOOLS

The **Solid Editing** tools can be selected from the **Home/Solid Editing** panel (Fig. 18.1).

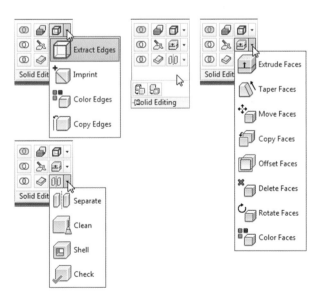

**Fig. 18.1** The **Home/Solid Editing** panel.

**Introduction to AutoCAD 2012. ISBN: 978-0-08-096947-3**

Examples of the results of using some of the **Solid Editing** tools are shown in this chapter. These tools are of value if the design of a 3D solid model is required to be changed (edited), although some have a value in constructing parts of 3D solids which cannot easily be constructed using other tools.

## FIRST EXAMPLE – EXTRUDE FACES TOOL (FIG. 18.3)

1. Set **ISOLINES** to 24.
2. In a **ViewCube/Right** view, construct a cylinder of radius **30** and height **30** (Fig. 18.2).

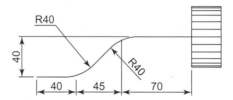

**Fig. 18.2** First example – **Extrude faces tool** – first stages.

3. In a **ViewCube/Front** view construct the pline (Fig. 18.2). Mirror the pline to the other end of the cylinder.
4. In a **ViewCube/Top** view move the pline to lie central to the cylinder.
5. Place the screen in a **ViewCube/Isometric view**.
6. *Click* the **Extrude faces** tool icon in the **Home/Solid Editing** panel (Fig. 18.1). The command line shows:

```
Command: _solidedit
Solids editing automatic checking: SOLIDCHECK=1
Enter a solids editing option [Face/Edge/Body/Undo/
 eXit] <eXit>: _face
Enter a face editing option
[Extrude/Move/Rotate/Offset/Taper/Delete/Copy/coLor/
 mAterial/Undo/eXit] <eXit>: _extrude
Select faces or [Undo/Remove]: pick the cylinder 2
 faces found
Select faces or [Undo/Remove/ALL]: enter r right-
 click
Remove faces or [Undo/Add/ALL]: right-click
Specify height of extrusion or [Path]: enter p
 (Path) right-click
Select extrusion path: pick the left-hand path pline
Solid validation started
```

```
Solid validation completed
Enter a face editing option [Extrude/Move/Rotate/
 Offset/Taper/Delete/Copy/coLor/mAterial/Undo/eXit]
 <eXit>: right-click
Command:
```

Original cylinder

After extruding faces along paths

**Fig. 18.3** First example – **Extrude faces** tool.

7. Repeat the operation using the pline at the other end of the cylinder as a path.
8. Add lights and a material and render the 3D model (Fig. 18.3).

## NOTE →

Note the prompt line which includes the statement **SOLIDCHECK=1**. If the variable **SOLIDCHECK** is set on (to 1) the prompt lines include the lines **SOLIDCHECK=1, Solid validation started**, and **Solid validation completed**. If set to 0 these two lines do not show.

Path – a pline

Extruded hexagon of height 1 unit

**Fig. 18.4** Second example – **Extrude faces** tool – pline for path.

## SECOND EXAMPLE – EXTRUDE FACES TOOL (FIG. 18.5)

1. Construct a hexagonal extrusion just **1** unit high in **ViewCube/ Top**.
2. In **ViewCube/Front** construct the curved pline in Fig. 18.4.
3. Back in the **Top** view, move the pline to lie central to the extrusion.
4. Place in **ViewCube/Isometric** and extrude the top face of the extrusion along the path of the curved pline.
5. Add lighting and a material to the model and render (Fig. 18.5).

## NOTE →

This example shows that a face of a 3D solid model can be extruded along any suitable path curve. If the polygon on which the extrusion had been based had been turned into a region, no extrusion could have taken place. The polygon had to be extruded to give a face to a 3D solid.

## THIRD EXAMPLE – MOVE FACES TOOL (FIG. 18.6)

1. Construct the 3D solid drawing shown in the left-hand drawing of Fig. 18.6 from three boxes which have been united using the **Union** tool.

**Fig. 18.5** Second example – **Extrude faces** tool.

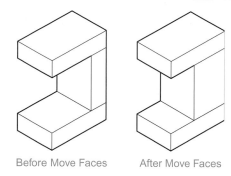

Before Move Faces  After Move Faces

**Fig. 18.6** Third example – **Solid, Move faces** tool.

2. *Click* on the **Move faces** tool in the **Home/Solid Editing** panel (see Fig. 18.1). The command line shows:

```
Command: _solidedit
[prompts] _face
Enter a face editing option
[prompts]: _move
Select faces or [Undo/Remove]: pick the model face
 4 faces found
Select faces or [Undo/Remove/ALL]: right-click
Specify a base point or displacement: pick
Specify a second point of displacement: pick
[further prompts]:
And the picked face is moved - right-hand drawing
 of Fig. 18.6.
```

## FOURTH EXAMPLE – OFFSET FACES (FIG. 18.7)

1. Construct the 3D solid drawing shown in the left-hand drawing of Fig. 18.9 from a hexagonal extrusion and a cylinder which have been united using the **Union** tool.
2. *Click* on the **Offset faces** tool icon in the **Home/Solid Editing** panel (Fig. 18.1). The command line shows:

```
Command: _solidedit
[prompts]: _face
[prompts]
[prompts]: _offset
Select faces or [Undo/Remove]: pick the bottom face
 of the 3D model 2 faces found
Select faces or [Undo/Remove/All]: enter r right-
 click
```

Select faces or [Undo/Remove/All]: *pick* highlighted
 faces other than the bottom face 2 faces found, 1
 removed
Select faces or [Undo/Remove/All]: *right-click*
Specify the offset distance: *enter* 30 *right-click*

**3.** Repeat, offsetting the upper face of the cylinder by **50** and the right-hand face of the lower extrusion by **15**.

The results are shown in Fig. 18.7.

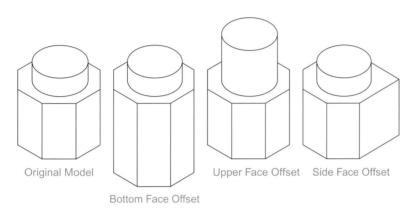

Original Model
Bottom Face Offset
Upper Face Offset   Side Face Offset

**Fig. 18.7** Fourth example – **Offset faces** tool.

## FIFTH EXAMPLE – TAPER FACES TOOL (FIG. 18.8)

**1.** Construct the 3D model as in the left-hand drawing of Fig. 18.8. Place in **ViewCube/Isometric**.
**2.** Call **Taper faces**. The command line shows:

Command: _solidedit
[prompts]: _face
[prompts]
[prompts]: _taper
Select faces or [Undo/Remove]: *pick* the upper face
 of the base 2 faces found
Select faces or [Undo/Remove/All]: *enter* r *right-*
 *click*
Select faces or [Undo/Remove/All]: *pick* highlighted
 faces other than the upper face 2 faces found, 1
 removed
Select faces or [Undo/Remove/All]: *right-click*
Specify the base point: *pick* a point on left-hand
 edge of the face

Specify another point along the axis of tapering: *pick* a point on the right-hand edge of the face
Specify the taper angle: *enter* 10 *right-click*

And the selected face tapers as indicated in the right-hand drawing of Fig. 18.8.

Before Taper Faces          After Taper Faces

**Fig. 18.8** Fifth example – **Taper faces** tool.

## SIXTH EXAMPLE – COPY FACES TOOL (FIG. 18.10)

**1.** Construct a 3D model to the sizes as given in Fig. 18.9.

**Fig. 18.9** Sixth example – **Copy Faces** tool – details of the 3D solid model.

**2.** *Click* on the **Copy faces** tool in the **Home/Solid Editing** panel (Fig. 18.1). The command line shows:

```
Command: _solidedit
[prompts]: _face
[prompts]
[prompts]: _copy
```

Select faces or [Undo/Remove]: *pick the upper face of the solid model 2 faces found*

Select faces or [Undo/Remove/All]: *enter r right-click*

Select faces or [Undo/Remove/All]: *pick highlighted face not to be copied 2 faces found, 1 removed*

Select faces or [Undo/Remove/All]: *right-click*

Specify a base point or displacement: *pick anywhere on the highlighted face*

Specify a second point of displacement: *pick a point some 50 units above the face*

**3.** Add lights and a material to the 3D model and its copied face and render (Fig. 18.10).

Before Copy Faces     After Copy Faces

**Fig. 18.10** Sixth example – **Copy faces** tool.

## SEVENTH EXAMPLE – COLOR FACES TOOL (FIG. 18.12)

**1.** Construct a 3D model of the wheel to the sizes as shown in Fig. 18.11.

**2.** *Click* the **Color faces** tool icon in the **Home/Solid Editing** panel (Fig. 18.1). The command line shows:

Command: _solidedit

[prompts]: _face

[prompts]

[prompts]: _color

Select faces or [Undo/Remove]: *pick the inner face of the wheel 2 faces found*

Select faces or [Undo/Remove/All]: *enter r right-click*

Select faces or [Undo/Remove/All]: *pick highlighted faces other than the required face 2 faces found, 1 removed*

**Fig. 18.11** Seventh example – **Color faces** tool – details of the 3D model.

```
Enter new color <ByLayer>: enter 1 (which is red)
 right-click
```

**3.** Add lights and a material to the edited 3D model and render (Fig. 18.12).

**Fig. 18.12** Seventh example – **Color faces** tool.

## EXAMPLES OF MORE 3D MODELS

The following 3D models can be constructed in the **acadiso3D.dwt** screen. The descriptions of the stages needed for their construction have been reduced from those given in earlier pages, in the hope that readers have already acquired a reasonable skill in the construction of such drawings.

# FIRST EXAMPLE (FIG. 18.14)

1. **Front** view. Construct the three extrusions for the back panel and the two extruding panels to the details given in Fig. 18.13.

**Fig. 18.13** First example – **3D models** – details of sizes and shapes.

2. **Top** view. Move the two panels to the front of the body and union the three extrusions. Construct the extrusions for the projecting parts holding the pin.
3. **Front** view. Move the two extrusions into position and union them to the back.
4. **Top** view. Construct two cylinders for the pin and its head.
5. **Top** view. Move the head to the pin and union the two cylinders.
6. **Front** view. Move the pin into its position in the holder. Add lights and materials.
7. **Isometric** view. Render. Adjust lighting and materials as necessary (Fig. 18.14).

**Fig. 18.14** First example – **3D models**.

# SECOND EXAMPLE (FIG. 18.16)

1. **Top.** Construct polyline outlines for the body extrusion and the solids of revolution for the two end parts. Extrude the body and subtract its hole and using the **Revolve** tool form the two end solids of revolution.
2. **Right.** Move the two solids of revolution into their correct positions relative to the body and union the three parts. Construct a cylinder for the hole through the model.
3. **Front.** Move the cylinder to its correct position and subtract from the model.

**Fig. 18.15** Second example – **3D models** dimensions.

4. **Top**. Add lighting and a material.
5. **Isometric**. Render (Fig. 18.16).

**Fig. 18.16** Second example – **3D models**.

## THIRD EXAMPLE (FIG. 18.18)

1. **Front**. Construct the three plines needed for the extrusions of each part of the model (details in Fig. 18.17). Extrude to the given heights. Subtract the hole from the **20** high extrusion.
2. **Top**. Move the **60** extrusion and the **10** extrusion into their correct positions relative to the **20** extrusion. With **Union** form a single 3D model from the three extrusions.
3. Add suitable lighting and a material to the model.
4. **Isometric**. Render (Fig. 18.18).

## FOURTH EXAMPLE (FIG. 18.19)

1. **Front**. Construct the polyline – left drawing in Fig. 18.19.
2. With the **Revolve** tool from the **Home/3D Modeling** panel construct a solid of revolution from the pline.

**Fig. 18.17** Third example – **3D models** – details of shapes and sizes.

**Fig. 18.18** Third example – **3D models.**

3. **Top**. Add suitable lighting a colored glass material.
4. **Isometric**. Render – right illustration in Fig. 18.19.

**Fig. 18.19** Fourth example – **3D models.**

# EXERCISES

Methods of constructing answers to the following exercises can be found in the free website:

**http://www.routledge.com/cw/yarwood**

1. Working to the shapes and dimensions as given in the orthographic projection in Fig. 18.20, construct the exploded 3D model as shown in Fig. 18.21. Once the model has been constructed add suitable lighting, choose materials, and render the model.

2. Working to the dimensions given in the orthographic projections of the three parts of the 3D model in Fig. 18.22, construct the assembly as shown in the rendered 3D model (Fig. 18.23). Add suitable lighting and materials, place in one of the isometric viewing positions, and render the model.

**Fig. 18.20** Exercise 1 – orthographic projection.

**Fig. 18.21** Exercise 1 – rendered 3D model.

**Fig. 18.22** Exercise 2 – details of shapes and sizes.

**Fig. 18.23** Exercise 2.

**3.** Construct the 3D model shown in the rendering in Fig. 18.24 from the details given in the parts drawing in Fig. 18.25.

**Fig. 18.24** Exercise 3.

**Fig. 18.25** Exercise 3 – the parts drawing.

**4.** A more difficult exercise: A rendered 3D model of the parts of an assembly is shown in Fig. 18.26. Working to the details given in the three orthographic projections in Figs 18.26, 18.27, and 18.28, construct the two parts of the 3D model, place them is suitable positions relative to each other, add lighting and materials, and render the model (Figs 18.29).

**Fig. 18.26** Exercise 4 – first orthographic projection.

Fillets are R2

**Fig. 18.27** Exercise 4 – third orthographic projections.

Holes Ø30

Hole Ø40

R60

R5

R40

10

50

55

50

27.5

92.5

152.5

15

Ø50

200

15

**Fig. 18.28** Exercise 4 – second orthographic projection.

**Fig. 18.29** Exercise 4.

# OTHER FEATURES OF 3D MODELING

## AIMS OF THIS CHAPTER

The aims of this chapter are:

**1.** To give a further example of placing raster images in an AutoCAD drawing.

**2.** To give examples of methods of printing or plotting not given in previous chapters.

**3.** To give examples of polygonal viewports.

## RASTER IMAGES IN AutoCAD DRAWINGS

### EXAMPLE – RASTER IMAGE IN A DRAWING (FIG. 19.5)

This example shows the raster file **Fig05.bmp** of the 3D model constructed to the details given in the drawing in Fig. 19.1.

**Fig. 19.1 Raster image in a drawing** – drawings into which file is to be inserted.

**Introduction to AutoCAD 2012. ISBN: 978-0-08-096947-3**

Raster images are graphics images in files with file names ending with the extensions *.bmp; *.pcx; *.tif and the like. The types of graphics files which can be inserted into AutoCAD drawings can be seen by first *clicking* on the **External References** palette icon in the **View/Palettes** panel (Fig.19.2).

**Fig. 19.2** Selecting **External Reference** palette from the **View/Palettes** panel.

**Fig. 19.3** The **External References** palette.

Then select **Attach Image** from the popup menu brought down with a *click* on the left-hand icon at the top of the palette. This brings the **Select Image File** dialog (Fig. 19.3), which will in turn bring the **Select Reference File** dialog on screen (Fig. 19.4).

In the dialog select the required raster file (in this example Fig05. bmp) and *click* the **Open** button. The **Attach Image** dialog appears showing the selected raster image. If satisfied *click* the **OK** button. The dialog disappears and the command line shows:

```
Command: _IMAGEATTACH
Specify insertion point <0,0>: pick
Base image size: Width: 1.000000, Height: 1.041958,
 Millimeters
Specify scale factor <1>: enter 60 right-click
Command:
```

And the image is attached on screen at the *picked* position.

## How to Produce a Raster Image

1. Construct the 3D model to the shapes and sizes given in Fig. 19.1 working in four layers each of a different color.
2. Place in the **ViewCube/Isometric** view.
3. Shade the 3D model in **Realistic** visual style.
4. **Zoom** the shaded model to a suitable size and press the **Print Scr** key of the keyboard.

**Fig. 19.4** Raster image in a drawing – the **Select Reference File** and **Attach Image** dialogs.

5. Open the Windows **Paint** application and *click* **Edit** in the menu bar, followed by another *click* on **Paste** in the drop-down menu. The whole AutoCAD screen which includes the shaded 3D assembled model appears.

6. *Click* the **Select** tool icon in the toolbar of **Paint** and window the 3D model. Then *click* **Copy** in the **Edit** drop-down menu.

7. *Click* **New** in the **File** drop-down menu, followed by a *click* on **No** in the warning window which appears.

8. *Click* **Paste** in the **Edit** drop-down menu. The shaded 3D model appears. *Click* **Save As...** from the **File** drop-down menu and save the bitmap to a suitable file name – in this example – **Fig05. bmp**.

9. Open the orthographic projection drawing in Fig. 19.1 in AutoCAD.

10. Following the details given on page 374 attach **Fig05.bmp** to the drawing at a suitable position (Fig. 19.5).

**Fig. 19.5** Example – **Raster image in a drawing**.

## JPGOUT

Another method of creating a raster image from a drawing in the AutoCAD widow is to use the command **jpgout**.

**1.** With the previous 3D model on screen, at the command line:
Command: *enter* jpgout *right-click*.
**2.** The **Create Raster File** dialog appears (Fig. 19.6). In the **File name** field *enter* Fig.06, followed by a *click* on the **OK** button.

**Fig. 19.6** First example – the **MODEL** button in the status bar.

The command line shows:

```
Command: jpgout
Command:
JPGOUT Select objects or <all objects and
  viewports>: pick top to the left-hand corner of
  model Specify opposite corner: pick bottom right-
  hand corner
16 found
Select objects or <all objects and viewports>:
Command:
```

**3.** The jpg raster image can be attached to a drawing using the same method as shown on page 374.

## NOTES →

1. It will normally be necessary to *enter* a scale in response to the prompt lines otherwise the raster image may appear very small on screen. If it does it can be zoomed anyway.
2. Place the image in position in the drawing area. In Fig. 19.5 the orthographic projections have been placed within a margin and a title block has been added.

## PRINTING/PLOTTING

Hardcopy (prints or plots on paper) from a variety of different types of AutoCAD drawings of 3D models can be obtained. Some of this variety has already been shown in Chapter 15.

## FIRST EXAMPLE – PRINTING/PLOTTING (FIG. 19.10)

If an attempt is made to print a multiple viewport screen in **Model Space** with all viewport drawings appearing in the plot, only the current viewport will be printed. To print or plot all viewports:

1. Open a 4-viewport screen of the assembled 3D model shown in the first example (Fig. 19.5).
2. Make a new layer **vports** of color **green**. Make this layer current.
3. *Click* the **MODEL** button in the status bar (Fig. 19.6). The **Page Setup Manager** dialog appears (Fig. 19.7). *Click* its **Modify...** button and the **Page Setup – Layout1** dialog appears (Fig. 19.8).
4. Make settings as shown and *click* the dialog's **OK** button; the **Page Setup Manager** dialog reappears showing the new settings. *Click* its **Close** button. The current viewport appears.
5. Erase the green outline and the viewport is erased.
6. At the command line:

```
Command: enter mv
MVIEW
Specify corner of viewport or
[ON/OFF/Fit/Shadeplot/Lock/Object/Polygonal/Restore/
  LAyer/2/3/4] <Fit>: enter r right-click
Enter viewport configuration name or [?] <*Active>:
  right-click
Specify first corner or [Fit] <Fit>: right-click
Command:
```

**Fig. 19.7** The **Page Setup Manager** dialog.

**Fig. 19.8** The **Page Setup – Layout1** dialog.

7. Turn layer **vports** off.
8. *Click* the **PAPER** button (note it changes from **MODEL**) and the current viewport changes to a model view. In each viewport in turn change the settings from the **3D Navigation** drop-down to **Front, Top, Right**, and **SW isometric**. *Click* the **MODEL** button. It changes to **PAPER** and the screen reverts to **Pspace**.
9. *Click* the **Plot** tool icon in the **Quick Access** bar (Fig. 19.9). A **Plot** dialog appears.

**Fig. 19.9** The **Plot** tool icon in the **Quick Access** toolbar.

**10.** Check in the dialog that the settings for the printer/plotter is correct and the paper size is also correct.

**11.** *Click* the **Preview** button. The full preview of the plot appears (Fig. 19.10).

**Fig. 19.10** First example – **Printing/plotting.**

**12.** *Right-click* anywhere in the drawing and *click* on **Plot** in the *right-click* menu which then appears.

**13.** The drawing plots (or prints).

## SECOND EXAMPLE – PRINTING/PLOTTING (FIG. 19.11)

**1.** Open the orthographic drawing with its raster image in Fig. 19.5.

**2.** While still in **Model Space** *click* the **Plot** tool icon. The **Plot** dialog appears. Check that the required printer/plotter and paper sizes have been chosen.

3. *Click* the **Preview** button.
4. If satisfied with the preview (Fig. 19.11), *right-click* and in the menu which appears *click* the name **Plot**. The drawing plots.

**Fig. 19.11** Second example – **Printing/Plotting.**

## THIRD EXAMPLE – PRINTING/PLOTTING (FIG. 19.12)

1. Open the 3D model drawing of the assembly shown in Fig. 19.10 in a single **ViewCube/Isometric** view.
2. While in **MSpace,** *click* the **Plot** tool icon. The **Plot** dialog appears.
3. Check that the plotter device and sheet sizes are correct. *Click* the **Preview** button.
4. If satisfied with the preview (Fig. 19.11) *right-click* and *click* on **Plot** in the menu which appears. The drawing plots.

## FOURTH EXAMPLE – PRINTING/PLOTTING (FIG. 19.13)

Fig. 19.13 shows a **Plot Preview** of the 3D solid model in Fig. 18.29 (see page 372).

The example to illustrate the construction of polygonal viewports is based upon **Exercise 4** of Chapter 18. When the 3D model for this exercise has been completed in **Model Space:**

1. Make a new layer **vports** of color **blue** and make it current.
2. Using the same methods as described for the first example of printing/plotting, produce a four-viewport screen of the model in **Pspace.**

**Fig. 19.12** Third example – **Printing/plotting**.

**Fig. 19.13** Fourth example – **Printing/Plotting**.

**3.** Erase the viewport with a *click* on its bounding line. The outline and its contents are erased.

**4.** *Click* the **Model** button. With a *click* in each viewport in turn and using the **ViewCube** settings, set viewports in **Front, Right, Top,** and **Isometric** views respectively.

**5.** **Zoom** each viewport to **All**.

**6.** *Click* the **Layout1** button to turn back to **PSpace**.

**7.** Enter **mv** at the command line, which shows:

```
Command: enter mv right-click
MVIEW
[prompts]: enter p (Polygonal) right-click
Specify start point: In the top right viewport pick
   one corner of a square
Specify next point or [Arc/Close/Length/Undo]: pick
   next corner for the square
Specify next point or [Arc/Close/Length/Undo]: pick
   next corner for the square
Specify next point or [Arc/Close/Length/Undo]: enter
   c (Close) right-click
Regenerating model
Command:
```

And a square viewport outline appears in the top right viewport
within which is a copy of the model.

**8.** Repeat in each of the viewports with different shapes of
polygonal viewport outlines (Fig. 19.14).

**9.** *Click* the **Model** button.

**10.** In each of the polygonal viewports make a different isometric
view. In the bottom-right polygonal viewport change the view
using the **3D Orbit** tool.

**11.** Turn the layer **vports** off. The viewport borders disappear.

**12.** *Click* the **Plot** icon. Make plot settings in the **Plot** dialog. *Click
on* the **Preview** button of the **Plot** dialog. The **Preview** appears
(Fig. 19.14).

**Fig. 19.14 Polygonal viewports** – plot preview – plot preview after **vports** layer is off.

## THE NAVIGATION WHEEL

The **Navigation Wheel** can be called from the **View/Navigate** panel. The reader is advised to experiment with the **Navigation Wheel** (Fig. 19.15).

**Fig. 19.15** The **Navigation Wheel**.

## THE MESH TOOLS

Fig. 19.16 shows a series of illustrations showing the actions of the **Mesh** tools and the three 3D tools **3dmove, 3dscale,** and **3drotate**.

The illustrations show:

1. A box constructed using the **Box** tool.
2. The box acted upon by the **Smooth Object** tool from the **Home/ Mesh** panel.
3. The box acted upon by the **Smooth Mesh** tool.
4. The box acted upon by the **Mesh Refine** tool.
5. The **Smooth refined** box acted upon by the **3dmove** tool.
6. The **Smooth Refined** box acted upon by the **3dscale** tool.
7. The **Smooth Refined** box acted upon by the **3drotate** tool.

**Fig. 19.16** Mesh: 3dmove, 3dscale, and 3drotate tools.

# EXERCISES

Methods of constructing answers to the following exercises can be found in the free website:

**http://www.routledge.com/cw/yarwood**

1. Working to the shapes and sizes given in Fig. 19.17, construct an assembled 3D model drawing of the spindle in its two holders, add lighting, and apply suitable materials and render (Fig. 19.18).

2. Fig. 19.19 shows a rendering of the model for Exercise 2 and Fig. 19.20 an orthographic projection giving shapes and sizes for the model. Construct the 3D model, add lighting, apply suitable materials, and render.

**Fig. 19.17** Exercise 1 – details of shapes and sizes.

**Fig. 19.18** Exercise 1.

**Fig. 19.19** Exercise 2.

**Fig. 19.20** Exercise 2 – orthographic projection.

**3.** Construct a 3D model drawing to the details given in Fig. 19.21. Add suitable lighting and apply a material, and then render as shown in Fig. 19.22.

**Fig. 19.21** Exercise 3 – **ViewCube/Isometric** view.

**Fig. 19.22** Exercise 3.

**4.** Construct an assembled 3D model drawing working to the details given in Fig. 19.23. When the 3D model drawing has been constructed disassemble the parts as shown in the given exploded 3D model (Fig. 19.24).

**Fig. 19.23** Exercise 4 – details of shapes and sizes.

**Fig. 19.24** Exercise 4 – an exploded rendered model.

**5.** Working to the details shown in Fig. 19.25, construct an assembled 3D model, with the parts in their correct positions relative to each other. Then separate the parts as shown in the 3D rendered model drawing Fig. 19.26. When the 3D model is complete add suitable lighting and materials and render the result.

**Fig. 19.25** Exercise 5 – details drawing.

**Fig. 19.26** Exercise 5 – exploded rendered view.

**6.** Working to the details shown in Fig. 19.27 construct a 3D model of the parts of the wheel with its handle. Two renderings of 3D models of the rotating handle are shown in Fig. 19.28, one with its parts assembled, the other with the parts in an exploded position relative to each other.

**Fig. 19.27** Exercise 6 – details drawing.

**Fig. 19.28** Exercise 6 – renderings.

7. Fig. 19.29 shows two pictures mounted in frames. Sections through the frames are given in Fig. 19.30. The pictures are the bitmaps (*.jpg format) shown in Fig. 19.31. Using either the given bitmaps or those of your own, construct two pictures in frames as shown.

**Fig. 19.29** Exercise 7.

**Fig. 19.30** Sections through the frames.

**Fig. 19.31** The two bitmaps.

# PART C

## INTERNET TOOLS AND HELP

C H A P T E R **20**

# INTERNET TOOLS AND HELP

## AIM OF THIS CHAPTER

The purpose of this chapter is to introduce the tools which are available in AutoCAD 2012 which make use of facilities available on the World Wide Web (WWW).

## EMAILING DRAWINGS

As with any other files that are composed of data, AutoCAD drawings can be sent by email as attachments. If a drawing needs to be password protected when emailing them, save the drawing with a password prior to attaching it to an email. To encrypt a drawing with a password, *click* **Tools** in the **Save Drawing As** dialog and from the popup list which appears *click* **Security Options...** (Fig. 20.1).

**Fig. 20.1** Selecting **Security Options** in the **Save Drawing As** dialog.

Then in the **Security Options** dialog which appears (Fig. 20.2) *enter* a password in the **Password or phrase to open this drawing** field, followed by a *click* on

**Introduction to AutoCAD 2012. ISBN: 978-0-08-096947-3**

**Fig. 20.2** *Entering* and confirming a password in the **Security Options** dialog.

the **OK** button. After *entering* a password *click* the **OK** button, and *enter* the password in the **Confirm Password** dialog which appears.

The drawing then cannot be opened until the password is *entered* in the **Password** dialog which appears when an attempt is made to open the drawing by the person receiving the email (Fig. 20.3).

**Fig. 20.3** The **Password** dialog appearing when a password encrypted drawing is about to be opened.

**Fig. 20.4** The **Publish to Web** tool in the **File** drop-down menu.

There are many reasons why drawings may require a password encapsulated in order to protect confidentiality of the contents of drawings.

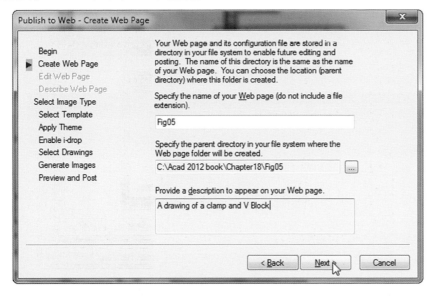

Fig. 20.5 The **Publish to Web – Create Web Page** dialog.

## CREATING A WEB PAGE (FIG. 20.5)

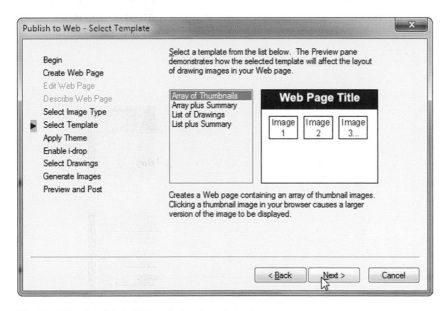

Fig. 20.6 The **Publish to Web – Select Template** dialog.

To create a web page which includes AutoCAD drawings, first *left-click* **Publish to Web...** in the **File** drop-down menu (Fig. 20.4).

A series of **Publish to Web** dialogs appear, some of which are shown here in Figs 20.5, 20.6, and 20.7.

**Fig. 20.7** The **Web Publishing – Windows Internet Explorer** page.

After making entries in the dialogs which come on screen after each **Next** button is *clicked*, the resulting web page such as that shown in Fig. 20.7 will be seen. *Double-click* in any of the thumbnail views in this web page and another page appears showing the selected drawing in full.

NOTE →

There is no icon for **eTransmit** in the ribbon panels.

THE ᴇTʀᴀɴꜱᴍɪᴛ TOOL

At the command line *enter* **etransmit**. The **Create Transmittal** dialog appears (Fig. 20.8). The transmittal shown in Fig. 20.8 is the drawing on screen at the time the transmittal was made plus a second drawing. Fill in details as necessary. The transmittal is transmitted in two parts.

**Fig. 20.8** The **Create Transmittal** dialog.

## HELP

Fig. 20.9 shows a method of getting help. In this example help on using the **Break** tool is required. *Enter* **Break** in the **Search** field, followed by a *click* on the **Click here to access the help** button (Fig. 20.9). The **AutoCAD 2012** Help page (Fig. 20.10) appears

**Fig. 20.9** Help for **Break**.

from which the operator can select what he/she considers to be the most appropriate response. In the web page screen, first, *click* the letter **B** in the **Command** list.

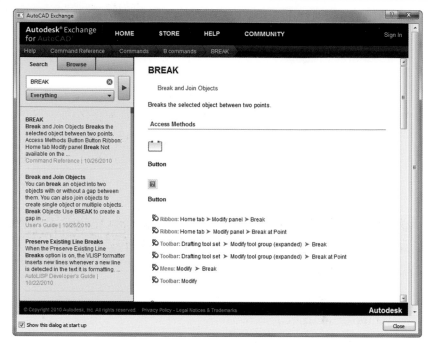

**Fig. 20.10** The Autodesk Exchange window showing Help for **Break**.

# DESIGN AND
# AutoCAD 2012

## AIMS OF THIS CHAPTER

The aims of this chapter are:

1. To describe reasons for using AutoCAD.
2. To describe methods of designing artifacts and the place of AutoCAD in the design process.
3. To list the system requirements for running AutoCAD 2012 software.
4. To list some of the enhancements in AutoCAD 2012.

## 10 REASONS FOR USING AutoCAD

1. A CAD software package such as AutoCAD 2012 can be used to produce any form of technical drawing.
2. Technical drawings can be produced much more speedily using AutoCAD than when working manually – probably as much as 10 times as quickly when used by skilled AutoCAD operators.
3. Drawing with AutoCAD is less tedious than drawing by hand – features such as hatching, lettering, adding notes, etc. are easier, quicker and indeed more accurate to construct.
4. Drawings or parts of drawings can be moved, copied, scaled, rotated, mirrored, and inserted into other drawings without having to redraw.
5. AutoCAD drawings can be saved to a file system without necessarily having to print the drawings. This can save the need for large paper drawing storage areas.

**Introduction to AutoCAD 2012. ISBN: 978-0-08-096947-3**

21

6. The same drawing or part of a drawing need never be drawn twice, because it can be copied or inserted into other drawings with ease. A basic rule when working with AutoCAD is:

*Never draw the same feature twice.*

7. New details can be added to drawings or be changed within drawings without having to mechanical erase the old detail.

8. Dimensions can be added to drawings with accuracy reducing the possibility of making errors.

9. Drawings can be plotted or printed to any scale without having to redraw.

10. Drawings can be exchanged between computers and/or emailed around the world without having to physically send them.

## THE PLACE OF AutoCAD 2012 IN DESIGNING

The contents of this book are only intended to help those who have a limited (or no) knowledge and skills of the construction of technical drawings using AutoCAD 2012. However, it needs to be recognized that the impact of modern computing on the methods of designing in industry has been immense. Such features as analysis of stresses, shear forces, bending forces, and the like can be carried out more quickly and accurately using computing methods. The storage of data connected with a design and the ability to recover the data speedily are carried out much more easily using computing methods than prior to the introduction of computing.

AutoCAD 2012 can play an important part in the design process because technical drawings of all types are necessary for achieving well designed artifacts whether it be an engineering component, a machine, a building, an electronics circuit or any other design project.

In particular, two-dimensional drawings (2D drawings) which can be constructed in AutoCAD 2012 are still of great value in modern industry. AutoCAD 2012 can also be used to produce excellent and accurate three-dimensional (3D) models, which can be rendered to produce photographic-like images of a suggested design. Although not dealt with in this book, data from 3D models constructed in AutoCAD 2012 can be taken for use in computer aided machining (CAM).

At all stages in the design process, either (or both) 2D or 3D drawings play an important part in aiding those engaged in designing to assist in assessing the results of their work at various stages. It is in the design process that drawings constructed in AutoCAD 2012 play an important part.

In the simplified design process chart shown in Fig. 21.1 an asterisk (*) has been shown against those features where the use of AutoCAD 2012 can be regarded as being of value.

## A DESIGN CHART (FIG. 21.1)

The simplified design chart in Fig. 21.1 shows the following features:

**Fig. 21.1** A simplified design chart.

**Design brief:** A design brief is a necessary feature of the design process. It can be in the form of a statement, but it is usually much more. A design brief can be a written report which not only includes a statement made of the problem which the design is assumed to be solving, but includes preliminary notes and drawings describing difficulties which may be encountered in solving the design and may include charts, drawings, costings etc. to emphasize some of the needs in solving the problem for which the design is being made.

**Research:** The need to research the various problems which may arise when designing is often much more demanding than the chart (Fig. 21.1) shows. For example the materials being used may require extensive research as to costing, stress analysis, electrical conductivity, difficulties in machining, or in constructional techniques and other such features.

**Ideas for solving the brief:** This is where technical, other drawings and sketches play an important part in designing. It is only after research that designers can ensure the brief will be fulfiled.

**Models:** These may be constructed models in materials representing the actual materials which have been chosen for the design but, in addition, 3D solid model drawings such as those which can be constructed in AutoCAD 2012, can be of value. Some models may also be made in the materials from which the final design is to be made so as to allow testing of the materials in the design situation.

**Chosen solution:** This is where the use of drawings constructed in AutoCAD 2012 are of great value. 2D and 3D drawings come into their own here. It is from such drawings that the final design will be manufactured.

**Realization:** The design is made. There may be a need to manufacture a number of designs in order to enable full assessment.

**Evaluation:** The manufactured design is tested in situations such as it is liable to be placed in use. Evaluation will include reports and notes which could include drawings with suggestions for amendments to the working drawings from which the design was realized.

## ENHANCEMENTS IN AUTOCAD 2012

AutoCAD 2012 contains many enhancements over previous releases, whether working in a **2D** or a **3D** workspace. Please note that not all the enhancements in AutoCAD 2012 are described in this introductory book. Among the more important enhancements are the following:

1. When first loaded, an **Initial Setup** dialog offers a **Welcome Screen** from which the operator can select from a variety of videos illustrating how different methods of drawing in both 2D and 3D can be used in AutoCAD 2012.
2. The **Ribbon** has been amended and brought up to date.
3. **Some new commands** have been introduced, particularly with the Array tool.
4. There is no **Array** dialog.
5. Some commands have been changed.
6. A number of new Set Variable introduced.
7. The **Materials Browser** has been revised and updated.
8. **Copy** tool includes a new **Array** option.
9. **Layers** have been enhanced.
10. **2D Array** has been changed including arrays along a path.
11. **Extend** tool can now be used with splines.

12. **Osnaps** have been enhanced.
13. **Mtext** has been enhanced.
14. A new tool **Offset Edge** has been introduced.
15. **UCS** has been enhanced.
16. **UCS** icon has been modified.   DESIGN AND AUTOCAD 2012

## SYSTEM REQUIREMENTS FOR RUNNING AutoCAD 2012

**Note:** there are two editions of AutoCAD 2012 – 32 bit and 64 bit editions.

**Operating system:** Windows XP Professional, Windows XP Professional (×64 Edition), Windows XP Home Edition, Windows 2000 or Windows Vista 32 bit, Windows Vista 64 bit, Windows 7.

**Microsoft Internet Explorer 7.0.**

**Processor:** Pentium III 800 MHz or equivalent

**Ram:** At least 128 MB.

**Monitor screen:** 1024 × 768 **VGA** with **True Color** as a minimum.

**Hard disk:** A minimum of 300 MB.

**Graphics card:** An AutoCAD certified graphics card. Details can be found on the web page **AutoCAD Certified Hardware XML Database**.

### NOTE →

Throughout this book illustrations show **Windows 7** dialogs, palettes, etc.

# LIST OF TOOLS

## INTRODUCTION

AutoCAD 2012 allows the use of over 1000 commands (or tools). A selection of the most commonly used commands, tools, are described in this appendix. Some of the commands described here have not been used in this book, because this book is an introductory text designed to initiate readers into the basic methods of using AutoCAD 2012. It is hoped the list will encourage readers to experiment with those tools not described in the book. The abbreviations, for the tools that have them, are included in brackets after the tool name. Tool names can be *entered* at the command line in upper or lower case.

A list of 2D commands is followed by a listing of 3D commands. Internet commands are described at the end of this listing. It must be remembered that not all of the tools available in AutoCAD 2012 are shown here.

The abbreviations for the commands can be found in the file **acad.pgp** from the folder: **C:\Autodesk\AutoCAD_2012_English_Win_32bit_SLD\x86\acad\en-us\Acad\Program Files\Root\UserDataCache\Support**

Not all of the commands have abbreviations.

## 2D COMMANDS

**About** – Brings the **About AutoCAD** bitmap on screen.
**Adcenter** (dc) – Brings the **DesignCenter** palette on screen.
**Align** (al) Aligns objects between chosen points.
**Appload** – Brings the **Load/Unload Applications** dialog to screen.
**Arc** (a) – Creates an arc.
**Area** – States in square units of the area selected from a number of points.
**Array** (ar) – Creates **Rectangular** or **Polar** arrays in 2D.
**Ase** – Brings the **dbConnect Manager** on screen.
**Attdef** – Brings the **Attribute Definition** dialog on screen.
**Attedit** – Allows editing of attributes from the Command line.
**Audit** – Checks and fixes any errors in a drawing.

**Autopublish** – Creates a **DWF** file for the drawing on screen.

**Bhatch** (h) – Brings the **Boundary Hatch** dialog on screen.

**Block** – Brings the **Block Definition** dialog on screen.

**Bmake** (b) – Brings the **Block Definition** dialog on screen.

**Bmpout** – Brings the **Create Raster File** dialog on screen.

**Boundary** (bo) – Brings the **Boundary Creation** dialog on screen.

**Break** (br) – Breaks an object into parts.

**Cal** – Calculates mathematical expressions.

**Chamfer** (cha) – Creates a chamfer between two entities.

**Chprop** (ch) – Brings the **Properties** window on screen.

**Circle** (c) – Creates a circle.

**Copy** (co) – Creates a single or multiple copies of selected entities.

**Copyclip** (Ctrl + C) – Copies a drawing, or part of a drawing for inserting into a document from another application.

**Copylink** – Forms a link between an AutoCAD drawing and its appearance in another application such as a word processing package.

**Copytolayer** – Copies objects from one layer to another.

**Customize** – Brings the **Customize** dialog to screen, allowing the customization of toolbars, palettes, etc.

**Dashboard** – Has the same action as **Ribbon**.

**Dashboardclose** – Closes the **Ribbon**.

**Ddattdef** (at) – Brings the **Attribute Definition** dialog to screen.

**Ddatte** (ate) – Edits individual attribute values.

**Ddcolor** (col) – Brings the **Select Color** dialog on screen.

**Ddedit** (ed) – The **Text Formatting** dialog box appears on selecting text.

**Ddim** (d) – Brings the **Dimension Style Manager** dialog box on screen.

**Ddinsert** (i) – Brings the **Insert** dialog on screen.

**Ddmodify** – Brings the **Properties** window on screen.

**Ddosnap** (os) – Brings the **Drafting Settings** dialog on screen.

**Ddptype** – Brings the **Point Style** dialog on screen.

**Ddrmodes** (rm) – Brings the **Drafting Settings** dialog on screen.

**Ddunits** (un) – Brings the **Drawing Units** dialog on screen.

**Ddview** (v) – Brings the **View Manager** on screen.

**Del** – Allows a file (or any file) to be deleted.

**Dgnexport** – Creates a **MicroStation V8 dgn** file from the drawing on screen.

**Dgnimport** – Allows a **MicroStation V8 dgn** file to be imported as an AutoCAD dwg file.

**Dim** – Starts a session of dimensioning.

**Dimension tools** – The **Dimension** toolbar contains the following tools – **Linear, Aligned, Arc Length, Ordinate, Radius, Jogged, Diameter, Angular, Quick Dimension, Baseline, Continue, Quick Leader, Tolerance, Center Mark, Dimension Edit, Dimension Edit Text, Update,** and **Dimension Style.**

**Dim1** – Allows the addition of a single addition of a dimension to a drawing.

**Dist** (di) – Measures the distance between two points in coordinate units.

**Distantlight** – Creates a distant light.

**Divide** (div) – Divides an entity into equal parts.

**Donut** (do) – Creates a donut.

**Dsviewer** – Brings the **Aerial View** window on screen.

**Dtext** (dt) – Creates dynamic text. Text appears in drawing area as it is entered.

**Dxbin** – Brings the **Select DXB File** dialog on screen.

**Dxfin** – Brings the **Select File** dialog on screen.

**Dxfout** – Brings the **Save Drawing As** dialog on screen.

**Ellipse** (el) – Creates an ellipse.

**Erase** (e) – Erases selected entities from a drawing.

**Exit** – Ends a drawing session and closes AutoCAD 2012

**Explode** (x) – Explodes a block or group into its various entities

**Explorer** – Brings the Windows **Explorer** on screen.

**Export** (exp) – Brings the **Export Data** dialog on screen.

**Extend** (ex) – To extend an entity to another.

**Fillet** (f) – Creates a fillet between two entities.

**Filter** – Brings the **Object Selection Filters** dialog on screen.

**Gradient** – Brings the **Hatch and Gradient** dialog on screen.

**Group** (g) – Brings the **Object Grouping** dialog on screen.

**Hatch** – Allows hatching by the *entry* responses to prompts.

**Hatchedit** (he) – Allows editing of associative hatching.

**Help** – Brings the **AutoCAD 2012 Help: User Documentation** dialog on screen.

**Hide** (hi) – To hide hidden lines in 3D models.

**Id** – Identifies a point on screen in coordinate units.

**Imageadjust** (iad) – Allows adjustment of images.

**Imageattach** (iat) – Brings the **Select Image File** dialog on screen.

**Imageclip** – Allows clipping of images.

**Import** – Brings the **Import File** dialog on screen.

**Insert** (i) – Brings the **Inert** dialog on screen.

**Insertobj** – Brings the **Insert Object** dialog on screen.

**Isoplane** (Ctrl/E) – Sets the isoplane when constructing an isometric drawing.

**Join** (j) – Joins lines which are in line with each other or arcs which are from the same center point.

**Laycur** – Changes layer of selected objects to current layer.

**Laydel** – Deletes and purges a layer with its contents.

**Layer** (la) – Brings the **Layer Properties Manager** dialog on screen.

**Layout** – Allows editing of layouts.

**Lengthen** (len) – Lengthens an entity on screen.

**Limits** – Sets the drawing limits in coordinate units.

**Line** (l) – Creates a line.

**Linetype** (lt) – Brings the **Linetype Manager** dialog on screen.

**List** (li) – Lists in a text window details of any entity or group of entities selected.

**Load** – Brings the **Select Shape File** dialog on screen.

**Ltscale** (lts) – Allows the linetype scale to be adjusted.

**Measure** (me) – Allows measured intervals to be placed along entities.

**Menu** – Brings the **Select Customization File** dialog on screen.

**Menuload** – Brings the **Load/Unload Customizations** dialog on screen.

**Mirror** (mi) – Creates an identical mirror image to selected entities.

**Mledit** – Brings the **Multiline Edit Tools** dialog on screen.

**Mline** (ml) – Creates mlines.

**Mlstyle** – Brings the **Multiline Styles** dialog on screen.

**Move** (m) – Allows selected entities to be moved.

**Mslide** – Brings the **Create Slide File** dialog on screen.

**Mspace** (ms) – When in PSpace changes to MSpace.

**Mtext** (mt or t) – Brings the **Multiline Text Editor** on screen.

**Mview** (mv) – To make settings of viewports in **Paper Space**.

**Mvsetup** – Allows drawing specifications to be set up.

**New** (Ctrl + N) – Brings the **Select template** dialog on screen.

**Notepad** – For editing files from the Windows **Notepad**.

**Offset** (o) – Offsets selected entity by a stated distance.

**Oops** – Cancels the effect of using **Erase**.

**Open** – Brings the **Select File** dialog on screen.

**Options** – Brings the **Options** dialog to screen.

**Ortho** – Allows **ortho** to be set ON/OFF.

**Osnap** (os) – Brings the **Drafting Settings** dialog to screen.

**Pagesetup** – Brings either the **Page Setup Manager** on screen.

**Pan** (p) – Drags a drawing in any direction.

**Pbrush** – Brings Windows **Paint** on screen.

**Pedit** (pe) – Allows editing of polylines. One of the options is **Multiple** allowing continuous editing of polylines without closing the command.

**Pline** (pl) – Creates a polyline.

**Plot** (Ctrl + P) – Brings the **Plot** dialog to screen.

**Point** (po) – Allows a point to be placed on screen.

**Polygon** (pol) – Creates a polygon.

**Polyline** (pl) – Creates a polyline.

**Preferences** (pr) – Brings the **Options** dialog on screen.

**Preview** (pre) – Brings the print/plot preview box on screen.

**Properties** – Brings the **Properties** palette on screen.

**Psfill** – Allows polylines to be filled with patterns.

**Psout** – Brings the **Create Postscript File** dialog on screen.

**Purge** (pu) – Purges unwanted data from a drawing before saving to file.

**Qsave** – Saves the drawing file to its current name in AutoCAD 2012.

**Quickcalc** (qc) Brings the **QUICKCALC** palette to screen.

**Quit** – Ends a drawing session and closes down AutoCAD 2012.

**Ray** – A construction line from a point.

**Recover** – Brings the **Select File** dialog on screen to allow recovery of selected drawings as necessary.

**Recoverall** – Repairs damaged drawing.

**Rectang** (rec) – Creates a pline rectangle.

**Redefine** – If an AutoCAD command name has been turned off by **Undefine**, **Redefine** turns the command name back on.

**Redo** – Cancels the last **Undo**.

**Redraw** (r) – Redraws the contents of the AutoCAD 2012 drawing area.

**Redrawall** (ra) – Redraws the whole of a drawing.

**Regen** (re) – Regenerates the contents of the AutoCAD 2012 drawing area.

**Regenall** (rea) – Regenerates the whole of a drawing.

**Region** (reg) – Creates a region from an area within a boundary.

**Rename** (ren) – Brings the **Rename** dialog on screen.

**Revcloud** – Forms a cloud-like outline around objects in a drawing to which attention needs to be drawn.

**Ribbon** – Brings the ribbon on screen.

**Ribbonclose** – Closes the ribbon.

**Save** (Ctrl + S) – Brings the **Save Drawing As** dialog box on screen.

**Saveas** – Brings the **Save Drawing As** dialog box on screen.

**Saveimg** – Brings the **Render Output File** dialog on screen.

**Scale** (sc) – Allows selected entities to be scaled in size – smaller or larger.

**Script** (scr) – Brings the **Select Script File** dialog on screen.

**Setvar** (set) – Can be used to bring a list of the settings of set variables into an AutoCAD Text window.

**Shape** – Inserts an already loaded shape into a drawing.

**Shell** – Allows MS-DOS commands to be entered.

**Sketch** – Allows freehand sketching.

**Solid** (so) – Creates a filled outline in triangular parts.

**Spell** (sp) – Brings the **Check Spelling** dialog on screen.

**Spline** (spl) – Creates a spline curve through selected points.

**Splinedit** (spe) – Allows the editing of a spline curve.

**Status** – Shows the status (particularly memory use) in a Text window.

**Stretch** (s) – Allows selected entities to be stretched.

**Style** (st) – Brings the **Text Styles** dialog on screen.

**Tablet** (ta) – Allows a tablet to be used with a pointing device.

**Tbconfig** – Brings the **Customize** dialog on screen to allow configuration of a toolbar.

**Text** – Allows text from the Command line to be entered into a drawing.

**Thickness** (th) – Sets the thickness for the **Elevation** command.

**Tilemode** – Allows settings to enable **Paper Space**.

**Tolerance** – Brings the **Geometric Tolerance** dialog on screen.

**Toolbar** (to) – Brings the **Customize User Interface** dialog on screen.

**Trim** (tr) – Allows entities to be trimmed up to other entities.

**Type** – Types the contents of a named file to screen.

**UCS** – Allows selection of **UCS** (user Coordinate System) facilities.

**Undefine** – Suppresses an AutoCAD command name.

**Undo** (u) (Ctrl + Z) – Undoes the last action of a tool.

**View** – Brings the **View** dialog on screen.

**Vplayer** – Controls the visibility of layers in **Paper Space**.

**Vports** – Brings the **Viewports** dialog on screen.

**Vslide** – Brings the **Select Slide File** dialog on screen.

**Wblock** (w) – Brings the **Create Drawing File** dialog on screen.

**Wipeout** – Forms a polygonal outline within which all crossed parts of objects are erased.

**Wmfin** – Brings the **Import WMF** dialog on screen.

**Wmfopts** – Brings the **WMF in Options** dialog on screen.

**Wmfout** – Brings the **Create WMF File** dialog on screen.

**Xattach** (xa) – Brings the **Select Reference File** dialog on screen.

**Xline** – Creates a construction line.

**Xref** (xr) – Brings the **Xref Manager** dialog on screen.

**Zoom** (z) – Brings the **Zoom** tool into action.

## 3D COMMANDS

**3darray** – Creates an array of 3D models in 3D space.

**3dcorbit** – Allows methods of manipulating 3D models on screen.

**3dface** (3f) – Creates a 3 or 4 sided 3D mesh behind which other features can be hidden.

**3ddistance** – Allows the controlling of the distance of 3D models from the operator.

**3ddwf** – brings up the **Export 3D DWF** dialog.

**3dfly** – Allows walkthroughs in any 3D plane.

**3dforbit** – Controls the viewing of 3D models without constraint.

**3dmesh** – Creates a 3D mesh in 3D space.

**3dmove** – Shows a 3D move icon.

**3dorbit** (3do) – Allows a continuous movement and other methods of manipulation of 3D models on screen.

**3dorbitctr** – Allows further and a variety of other methods of manipulation of 3D models on screen.

**3dpan** – Allows the panning of 3D models vertically and horizontally on screen.

**3drotate** – displays a 3D rotate icon.

**3dsin** – Brings the **3D Studio File Import** dialog on screen.

**3dsout** – Brings the **3D Studio Output File** dialog on screen.

**3dwalk** – Starts walk mode in 3D.

**Align** – Allows selected entities to be aligned to selected points in 3D space.

**Ameconvert** – Converts AME solid models (from Release 12) into AutoCAD 2000 solid models.

**anipath** – Opens the **Motion Path Animation** dialog.

**Box** – Creates a 3D solid box.

**Cone** – Creates a 3D model of a cone.

**convertoldlights** – Converts lighting from previous releases to AutoCAD 2009 lighting.

**convertoldmaterials** – Converts materials from previous releases to AutoCAD 2009 materials.

**convtosolid** – Converts plines and circles with thickness to 3D solids.

**convtosurface** – Converts objects to surfaces.

**Cylinder** – Creates a 3D cylinder.

**Dducs** (uc) – Brings the **UCS** dialog on screen.

**Edgesurf** – Creates a 3D mesh surface from four adjoining edges.

**Extrude** (ext) Extrudes a closed polyline.

**Flatshot** – Brings the **Flatshot** dialog to screen.

**Freepoint** – Point light created without settings.

**Freespot** – Spot light created without settings.

**Helix** – Construct a helix.

**Interfere** – Creates an interference solid from selection of several solids.

**Intersect** (in) – Creates an intersection solid from a group of solids.

**Light** – Enables different forms of lighting to be placed in a scene.

**Lightlist** – Opens the **Lights in Model** palette.

**Loft** – Activates the **Loft** command.

**Materials** – Opens the **Materials** palette.

**Matlib** – Outdated instruction.

**Mirror3d** – Mirrors 3D models in 3D space in selected directions.

**Mview** (mv) – When in PSpace brings in MSpace objects.

**Pface** – Allows the construction of a 3D mesh through a number of selected vertices.

**Plan** – Allows a drawing in 3D space to be seen in plan (UCS World).

**Planesurf** – Creates a planar surface.

**Pointlight** – Allows a point light to be created.

**Pspace** (ps) – Changes MSpace to PSpace.

**Pyramid** – Creates a pyramid.

**Renderpresets** – Opens the **Render Presets Manager** dialog.

**Renderwin** – Opens the **Render** window.

**Revolve** (rev) – Forms a solid of revolution from outlines.

**Revsurf** – Creates a solid of revolution from a pline.

**Rmat** – Brings the **Materials** palette on screen.

**Rpref** (rpr) – Opens the **Advanced Render Settings** palette.

**Section** (sec) – Creates a section plane in a 3D model.

**Shade** (sha) – Shades a selected 3D model.

**Slice** (sl) – Allows a 3D model to be cut into several parts.

**Solprof** – Creates a profile from a 3D solid model drawing.

**Sphere** – Creates a 3D solid model sphere.

**Spotlight** – Creates a spotlight.

**Stlout** – Saves a 3D model drawing in ASCII or binary format.

**Sunproperties** – Opens the **Sun Properties** palette.

**Torus** (tor) – Allows a 3D torus to be created.

**Ucs** – Allows settings of the UCS plane.

**-render** – Can be used to make rendering settings from the command line. Note the hyphen (-) must precede **render**.

**Sweep** – Creates a 3D model from a 2D outline along a path.

**Tabsurf** – Creates a 3D solid from an outline and a direction vector.

**Ucs** – Allows settings of the UCS plane.

**Union** (uni) – Unites 3D solids into a single solid.

**View** – Creates view settings for 3D models.

**Visualstyles** – opens the **Visual Styles Manager** palette.

**Vpoint** – Allows viewing positions to be set from $x,y,z$ entries.

**Vports** – Brings the **Viewports** dialog on screen.

**Wedge** (we) – Creates a 3D solid in the shape of a wedge.

**Xedges** – creates a 3D wireframe for a 3D solid.

## INTERNET TOOLS

**Etransmit** – Brings the **Create Transmittal** dialog to screen.

**Publish** – Brings the **Publish** dialog to screen.

APPENDIX **B**

# SOME SET VARIABLES

## INTRODUCTION

AutoCAD 2012 is controlled by a large number of set variables (over 770 in number), the settings of many of which are determined when making entries in dialogs. Some are automatically set with *clicks* on tool icons. Others have to be set at the command line. Some are read-only variables and depend upon the configuration of AutoCAD 2012 when it was originally loaded into a computer (default values). Only a limited number of the variables are shown here.

A list of those set variables follows which are of interest in that they often require setting by *entering* figures or letters at the command line. To set a variable, enter its name at the command line and respond to the prompts which arise.

To see all set variables, *enter***set (or setvar)** at the command line:

```
Command: enter set right-click
SETVAR Enter variable name or ?: enter?
Enter variable name to list <*>: right-click
```

And an **AutoCAD Text Window** opens showing a list of the first of the set variables. To continue with the list press the **Return** key when prompted and at each press of the **Return** key, another window opens.

To see the settings needed for a set variable *enter* the name of the variable at the command line, followed by pressing the **F1** key which brings up a **Help** screen, *clicking* the search tab, followed by *entering* set variables in the **Ask** field. From the list then displayed the various settings of all set variables can be read.

## SOME OF THE SET VARIABLES

**ANGDIR** – Sets angle direction. **0** counterclockwise;
**1** clockwise.
**APERTURE** – Sets size of pick box in pixels.

AUTODWFPUBLISH – Sets **Autopublish** on or off.

BLIPMODE – Set to **1** marker blips show; set to **0** no blips.

COMMANDLINE – Opens the command line palette.

COMMANDLINEHIDE – Closes the command line palette.

COPYMODE – Sets whether **Copy** repeats.

## NOTES →

DIM variables – There are over 70 variables for setting dimensioning, but most are in any case set in the **Dimension Styles** dialog or as dimensioning proceeds. However, one series of the **Dim** variables may be of interest.

DIMBLK1 – Operator's arrowhead for first end of line.

DIMBLK2 – Operator's arrowhead for other end of line.

DMBLOCK – Sets a name for the block drawn for an operator's own arrowheads. These are drawn in unit sizes and saved as required.

DRAG1 – Sets regeneration drag sampling. Initial value is 10.

DRAG2 – Sets fast dragging regeneration rate. Initial value is 25.

DRAGMODE – Set to **0** no dragging; set to **1** dragging on; set to **2** automatic dragging.

FILEDIA – Set to **0** disables **Open** and **Save As** dialogs; set to **1** enables these dialogs.

FILLMODE – Set to **0** hatched areas are filled with hatching. Set to **0** hatched areas are not filled. Set to **0** and plines are not filled.

GRIPS – set to **1** and grips show. Set to **0** and grips do not show.

LIGHTINGUNITS – set to **1** (international) or **2** (USA) for photometric lighting to function.

MBUTTONPAN – Set to **0** no *right-click* menu with the Intellimouse. Set to **1** Intellimouse *right-click* menu on.

MIRRTEXT – Set to **0** text direction is retained; set to **1** text is mirrored.

NAVVCUBE – Sets the **ViewCube** on/off.

NAVVCUBELOCATION – Controls the position of the **ViewCube** between top-right (**0**) and bottom-left (**3**).

NAVVCUBEOPACITY – Controls the opacity of the **ViewCube** from **0** (hidden) to **100** (dark).

NAVVCUBESIZE – controls the size of the **ViewCube** between **0** (small to **2** (large).

PELLIPSE – Set to **0** creates true ellipses; set to **1** polyline ellipses.

PERSPECTIVE – Set to **0** places the drawing area into parallel projection. Set to **1** places the drawing area into perspective projection.

→ NOTES (CONTINUED)

**PICKBOX** – Sets selection pick box height in pixels.

**PICKDRAG** – Set to **0** selection windows picked by two corners; set to **1** selection windows are dragged from corner to corner.

**RASTERPREVIEW** – Set to **0** raster preview images not created with drawing. Set to **1** preview image created.

**SHORTCUTMENU** – For controlling how *right-click* menus show:

**0** all disabled; **1** default menus only; **2** edit mode menus; **4** command mode menus; **8** command mode menus when options are currently available. Adding the figures enables more than one option.

**SURFTAB1** – Sets mesh density in the M direction for surfaces generated by the **Surfaces** tools.

**SURFTAB2** – Sets mesh density in the N direction for surfaces generated by the **Surfaces** tools.

**TEXTFILL** – Set to **0** True Type text shows as outlines only; set True Type text is filled.

**TILEMODE** – Set to **0** Paperspace enabled; set to **1** tiled viewports in Modelspace.

**TOOLTIPS** – Set to **0** no tool tips; set to **1** tool tips enabled.

**TPSTATE** – Set to **0** and the Tool Palettes window is inactive. Set to **1** and the Tool Palettes window is active.

**TRIMMODE** – Set to **0** edges not trimmed when **Chamfer** and **Fillet** are used; set to **1** edges are trimmed.

**UCSFOLLOW** – Set to **0** new UCS settings do not take effect; set to **1** UCS settings follow requested settings.

**UCSICON** – Set **OFF** UCS icon does not show; set to **ON** and it is visible.

# APPENDIX **C**

# RIBBON TOOL PANEL ICONS

## INTRODUCTION

The ribbon panels shown here are those which include tools described in the earlier chapters of this book. Where flyouts from a panel include tools icons, the flyouts have been shown. Flyouts appear when an arrow at the bottom of the panel name is *clicked*. Where the names of tool icons have been included in the panels, the names have not been added to the illustrations as labels have deemed to be unnecessary.

## DRAFTING & ANNOTATION RIBBON

**Fig. C.1** The **Home/Draw** panel.

Fig. C.2 The **Home/Draw** flyout.

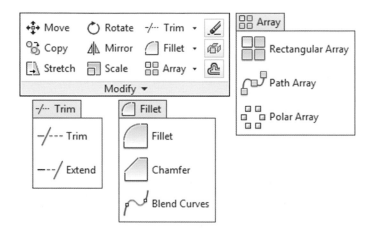

Fig. C.3 The **Home/Modify** panel.

Fig. C.4 The **Home/Draw** panel flyout.

Fig. C.5 The **Home/Layers** panel with the **Layers** drop-down menu.

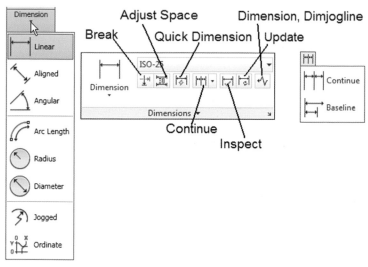

**Fig. C.6** The **Annotate/Dimensions** panel.

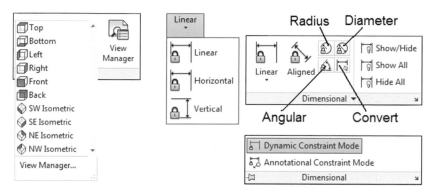

**Fig. C.7** The **View/Views** panel.

**Fig. C.8** The **Parametric/Dimensions** panel.

**Fig. C.9** The **View/Navigate 2D** panel.

**Fig. C.10** The **View/Palettes** panel.

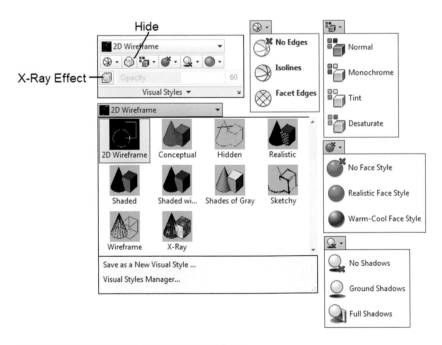

**Fig. C.11** The **View/Visual Styles** panel and flyout.

**Fig. C.12** The **View/Viewports** panel.

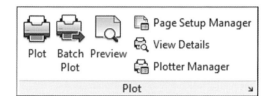

**Fig. C.13** The **Output/Plot** panel.

**Fig. C.14** The **Output/Export to DWF/PDF** panel.

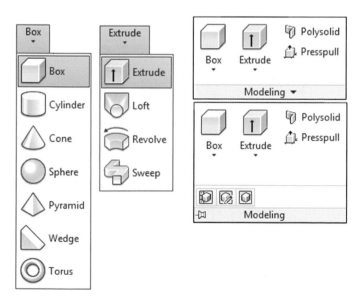

**Fig. C.15** The **Home/Modeling** panel and its flyout.

**Fig. C.16** The **Home/Solid Editing** panel.

**Fig. C.17** The **Home/Modify** panel.

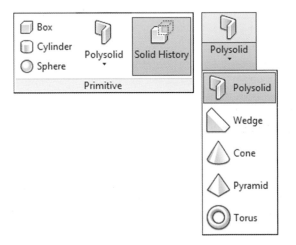

**Fig. C.18** The **Home/Modify** panel flyout.

**Fig. C.19** The **Solid/Primitive** panel.

**Fig. C.20** The **Solid/Solid** panel.

**Fig. C.21** The **Solid/Boolean** panel.

**Fig. C.22** The **Solid/Editing** panel.

**Fig. C.23** The **Render/Lights** panel and its flyout.

**Fig. C.24** The **Render/Materials** panel.

**Fig. C.25** The **Render/Render** panel.

**Fig. C.26** the **Render/Render** flyout.

# 3D VIEWS

## INTRODUCTION

## FROM THE VIEWS DROP-DOWN MENU IF SHOWING IN THE MENU BAR

There are a number of methods of setting the positions of 3D views, some of which have not been shown within the contents of this book. When setting a 3D view, any of the methods shown in this appendix can be used.

**Fig. D.1** The **View/3D Views** drop-down menu.

# FROM THE 3D MODELING RIBBON/ VIEW PANEL

**Fig. D.2** The **View/Views** panel drop-down menu.

# FROM THE VIEW MANAGER DIALOG

Either *left-click* on **View Manager** at the bottom of the **View/View** panel, or *enter* **view** at the command line to bring up the **View Manager** dialog.

**Fig D.3** The **View Manager** dialog.

# FROM THE VIEWCUBE

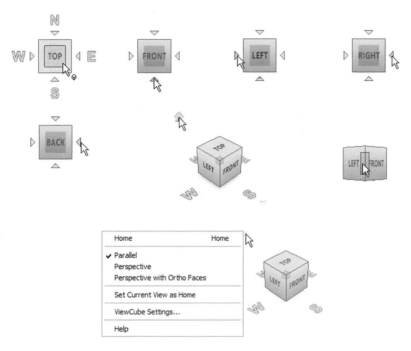

**Fig. D.4** Some settings of the **ViewCube**.

# FROM 3D BASICS LAYERS AND HOME/
# LAYERS & VIEWS/3D NAVIGATION

**Fig. D.5** *Left-click* on **3D Navigation**.

# FROM THE VIEWPORTS DIALOG

**Fig. D.6** Selecting views from the **Visual Style** popup in the **Viewports** dialog.

# FROM THE VIEWPOINT PRESETS DIALOG

**Fig. D.7** The **Viewpoint Presets** dialog.

# USING THE 3D ORBIT TOOL

**Fig. D.8** Selecting the 3D Orbit tool from the **View/Navigate** panel.

## FROM THE UCS

First set **ucsfollow** to **1**.

```
Command: enter ucsfollow right-click
Enter new value for UCSFOLLOW <0>: enter 1 right-
 click
Command:
```

**THEN:**

```
Command: enter ucs right-click
Current ucs name: *WORLD*
Specify origin of UCS or [Face/NAmed/OBject/
 Previous/View/World/X/Y/Z/ZAxis]
<World>: enter g right-click
Enter an option [Top/Bottom/Front/BAck/Left/
 Right]<Top>:
```

**THEN:**

*enter* the capital letter of the required 3D view, followed by a *right-click*.

# KEYBOARD SHORTCUTS

## KEYBOARD SHORTCUTS

| | |
|---|---|
| Ctrl + A: | Selects everything on screen. |
| Ctrl + C: | Calls the Copy command. |
| Ctrl + N: | Opens the Select template dialog. |
| Ctrl + O: | Opens the Select File dialog. |
| Ctrl + P: | Opens the Plot dialog. |
| Ctrl + Q: | Closes the AutoCAD window. |
| Ctrl + S: | Saves drawing on screen. |
| Ctrl + V: | Pastes from Clipboard into window. |
| Ctrl + X: | Calls the Cut command. |
| Ctrl + Y: | Calls the Redo command. |
| Ctrl + Z: | Undoes the last Plot operation. |
| Ctrl + Shift + C: | Calls the Copy command with Base point. |
| Ctrl + Shift + V: | Pastes a block into the window. |
| Ctrl + 5: | Saves the drawing in screen. |
| Ctrl + Shift + 5: | Opens the Save Drawing As dialog. |
| Ctrl + 9: | Toggles the Command palette on/off. |
| Del: | Deletes a selected object. |
| F1: | Brings the Help window on screen. |
| F2: | Brings the Text window on screen. |
| F3: | Toggles Object Snap on/off. |
| F4: | Toggles 3D Object Snap on/off. |

| F5: | Toggles between isoplanes. |
|-----|---------------------------|
| F6: | Toggles Dynamic UCS on/off. |
| F7: | Toggles Grid on/off. |
| F8: | Toggles Ortho on/off. |
| F9: | Toggles Snap on/off. |
| F10: | Toggles Polar Tracking on/off. |
| F11: | Toggles Object Tracking on/off. |

# Index